Springer Series in Statistics

Advisors:
P. Bickel, P. Diggle, S. Fienberg, K. Krickeberg,
I. Olkin, N. Wermuth, S. Zeger

Springer
New York
Berlin
Heidelberg
Barcelona
Hong Kong
London
Milan
Paris
Singapore
Tokyo

Springer Series in Statistics

Andersen/Borgan/Gill/Keiding: Statistical Models Based on Counting Processes.
Berger: Statistical Decision Theory and Bayesian Analysis, 2nd edition.
Bolfarine/Zacks: Prediction Theory for Finite Populations.
Borg/Groenen: Modern Multidimensional Scaling: Theory and Applications
Brockwell/Davis: Time Series: Theory and Methods, 2nd edition.
Chen/Shao/Ibrahim: Monte Carlo Methods in Bayesian Computation.
Efromovich: Nonparametric Curve Estimation: Methods, Theory, and Applications.
Fahrmeir/Tutz: Multivariate Statistical Modelling Based on Generalized Linear Models.
Farebrother: Fitting Linear Relationships: A History of the Calculus of Observations 1750-1900.
Federer: Statistical Design and Analysis for Intercropping Experiments, Volume I: Two Crops.
Federer: Statistical Design and Analysis for Intercropping Experiments, Volume II: Three or More Crops.
Fienberg/Hoaglin/Kruskal/Tanur (Eds.): A Statistical Model: Frederick Mosteller's Contributions to Statistics, Science and Public Policy.
Fisher/Sen: The Collected Works of Wassily Hoeffding.
Good: Permutation Tests: A Practical Guide to Resampling Methods for Testing Hypotheses, 2nd edition.
Gouriéroux: ARCH Models and Financial Applications.
Grandell: Aspects of Risk Theory.
Haberman: Advanced Statistics, Volume I: Description of Populations.
Hall: The Bootstrap and Edgeworth Expansion.
Härdle: Smoothing Techniques: With Implementation in S.
Hart: Nonparametric Smoothing and Lack-of-Fit Tests.
Hartigan: Bayes Theory.
Hedayat/Sloane/Stufken: Orthogonal Arrays: Theory and Applications.
Heyde: Quasi-Likelihood and its Application: A General Approach to Optimal Parameter Estimation.
Huet/Bouvier/Gruet/Jolivet: Statistical Tools for Nonlinear Regression: A Practical Guide with S-PLUS Examples.
Kolen/Brennan: Test Equating: Methods and Practices.
Kotz/Johnson (Eds.): Breakthroughs in Statistics Volume I.
Kotz/Johnson (Eds.): Breakthroughs in Statistics Volume II.
Kotz/Johnson (Eds.): Breakthroughs in Statistics Volume III.
Küchler/Sørensen: Exponential Families of Stochastic Processes.
Le Cam: Asymptotic Methods in Statistical Decision Theory.
Le Cam/Yang: Asymptotics in Statistics: Some Basic Concepts, 2nd edition.
Longford: Models for Uncertainty in Educational Testing.
Miller, Jr.: Simultaneous Statistical Inference, 2nd edition.
Mosteller/Wallace: Applied Bayesian and Classical Inference: The Case of the Federalist Papers.
Parzen/Tanabe/Kitagawa: Selected Papers of Hirotugu Akaike.
Politis/Romano/Wolf: Subsampling.

(continued after index)

Lucien Le Cam
Grace Lo Yang

Asymptotics in Statistics

Some Basic Concepts

Second Edition

Lucien Le Cam
Department of Statistics
University of California, Berkeley
Berkeley, CA 94720
USA

Grace Lo Yang
Department of Mathematics
University of Maryland
College Park, MD 20742
USA
gly@math.umd.edu

Library of Congress Cataloging-in-Publication Data
Le Cam, Lucien M. (Lucien Marie), 1924–
Asymptotics in statistics : some basic concepts / Lucien Le Cam, Grace Lo Yang.
—2nd ed.
p. cm. — (Springer series in statistics)
Includes bibliographical references and index.
ISBN 0-387-95036-2 (alk. paper)
1. Mathematical statistics—Asymptotic theory. I. Yang, Grace Lo. II. Title.
III. Series.
QA276 .L336 2000
519.5—dc21 00-030759

Printed on acid-free paper.

Production managed by Michael Koy; manufacturing supervised by Jerome Basma.
Photocomposed copy prepared by the authors using LaTeX.
Printed and bound by Edwards Brothers, Inc., Ann Arbor, MI.
Printed in the United States of America.

9 8 7 6 5 4 3 2 1

ISBN 0-387-95036-2 Springer-Verlag New York Berlin Heidelberg SPIN 10524420

Dedicatory Note

On April 25, 2000, as the last editorial touches were added to our book, Lucien Le Cam passed away. At Berkeley, he worked without stopping on the final revision of our book up until the time when he was hospitalized, just four days before his death.

His passing is a great loss to the statistical profession. He was a fundamental thinker in mathematical statistics and a principal architect of the modern asymptotic theory of statistics, following and extending the path of Neyman and Wald. Among the numerous seminal concepts that he introduced, the Le Cam distance between experiments provides a coherent statistical theory that links asymptotics and decision theory.

I was privileged and extremely fortunate to be his student and later his collaborator on several projects, including this book. Statisticians all know that Professor Le Cam was a brilliant mathematician. His students, colleagues, and friends also know of his kindness, generosity, integrity and patience.

This book is dedicated to the memory of Professor Le Cam. We feel his loss keenly. He will be greatly missed.

Grace Lo Yang
College Park, Maryland

May 2000

Preface to the Second Edition

This is a second edition, "reviewed and enlarged" as the French used to say. The 1990 edition was generally well received, but complaints about the conciseness of the presentation were heard. We have tried to accommodate such concerns, aiming at being understood by a "good" second-year graduate student.

The first edition contained some misleading typos. They have been removed, but we cannot guarantee that the present version is totally free of misprints.

Among substantial changes, we have introduced a Chapter 4 on gaussian and Poisson experiments. The gaussian ones are used everywhere because of mathematical tractability. Poisson experiments are less tractable and less studied, but they are probably more important. They will loom large in the new century.

The proof of asymptotic sufficiency of the "centerings" of Chapter 6 has been reworked and simplified. So has Chapter 7, Section 4. Section 5 of the same chapter has been augmented by a derivation of the Cramér-Rao lower bound imitated from Barankin [1949].

We thought of adding material on the von Mises differentiable functions and similar entities; we did not. The reader can refer to the recent book by van der Vaart [1998]. Similarly, we did not cover the use of "empirical processes" even though, following Dudley's paper of 1978, they have been found most valuable. A treatment of that material is expected to appear in a book by Sara van de Geer. Also, we did not include material due to David Donoho, Iain Johnstone, and their school. We found ourselves unprepared to write a distillate of the material. We did touch briefly on "nonparametrics," but not on "semiparametrics." This is because we feel that the semiparametric situation has not yet been properly structured.

We hope that the reader will find this book interesting and challenging, in spite of its shortcomings.

The material was typed in LaTeX form by the authors themselves, borrowing liberally from the 1990 script by Chris Bush. It was reviewed anonymously by distinguished colleagues. We thank them for their kind encouragement. Very special thanks are due to Professor David Pollard who took time out of a busy schedule to give us a long list of suggestions. We did not follow them all, but we at least made attempts.

We wish also to thank the staff of Springer-Verlag for their help, in particular editor John Kimmel, who tried to make us work with all deliberate speed. Thanks are due to Paul Smith, Te-Ching Chen and Ju-Yi-Yen, who helped with the last-minute editorial corrections.

Lucien Le Cam
Berkeley, California
Grace Lo Yang
College Park, Maryland

February 2000

Preface to the First Edition

In the summer of 1968 one of the present authors (LLC) had the pleasure of giving a sequence of lectures at the University of Montreal. Lecture notes were collected and written out by Drs. Catherine Doléans, Jean Haezendonck and Roch Roy. They were published in French by the Presses of the University of Montreal as part of their series of *Séminaires de Mathématiques Supérieures.* Twenty years later it was decided that a Chinese translation could be useful, but upon prodding by Professor Shanti Gupta at Purdue we concluded that the notes should be updated and rewritten in English and in Chinese. The present volume is the result of that effort.

We have preserved the general outline of the lecture notes, but we have deleted obsolete material and sketched some of the results acquired during the past twenty years. This means that while the original notes concentrated on the LAN situation we have included here some results of Jeganathan and others on the LAMN case. Also included are versions of the Hájek-Le Cam asymptotic minimax and convolution theorems with some of their implications. We have not attempted to give complete coverage of the subject and have often stated theorems without indicating their proofs.

What we have attempted to do is to present a few concepts and tools in an elementary manner refering the reader to the general literature for further information. We hope that this will provide the reader with a way of thinking about asymptotic problems in statistics that is somewhat more coherent than the traditional reliance upon maximum likelihood.

We wish to extend our thanks to the Presses of the University of Montreal for the permission to reuse some copyrighted material and to Springer-Verlag for the production of the present volume.

We also extend all our thanks to Professor Kai-Tai Fang whose efforts with Science Press are very much appreciated.

The English version of the manuscript was typed at Berkeley by Ms. Chris Bush whose patience and skill never cease to amaze us.

As Chris can attest, producing the typescript was no simple task. We were fortunate to have the help of Ruediger Gebauer and Susan Gordon at Springer-Verlag. We are very grateful for the assistance of Mr. Jian-Lun Xu in the preparation of the Chinese version.

Lucien Le Cam
Berkeley, California
Grace Lo Yang
College Park, Maryland

October 1989

Contents

1

Introduction

In this volume we describe a few concepts and tools that we have found useful in thinking about asymptotic problems in statistics. They revolve largely around the idea of approximating a family of measures, say, $\mathcal{E} = \{P_\theta; \theta \in \Theta\}$ by other families, say, $\mathcal{F} = \{Q_\theta; \theta \in \Theta\}$ that may be better known or more tractable.

For example, contemplate a situation where the statistician observes a large number of independent, identically distributed variables $X_1, X_2, \ldots, X_n$ that have a common Cauchy distribution with density

$$f(x, \theta) = \frac{1}{\pi} \frac{1}{1 + (x - \theta)^2}$$

on the line.

Let $P_{\theta,n}$ be the joint distribution of $X_1, \ldots, X_n$. Let Z_n be another variable that has a gaussian distribution $G_{\theta,n}$ with expectation θ and variance $2/n$ on the real line.

The theory expounded in later chapters says that, for n large, the two families $\mathcal{E}_n = \{P_{\theta,n}; \theta \in \Re\}$ and $\mathcal{F}_n = \{G_{\theta,n}; \theta \in \Re\}$ are, for most statistical purposes, very close to each other.

For another example, suppose that $Y_1, Y_2, \ldots Y_n$ are independent with a common density $[1 - |x - \theta|]^+$ on the line. Let $Q_{\theta,n}$ be their joint distribution and let $H_{\theta,n}$ be gaussian with mean θ and variance $1/(n \log n)$. Then, for n large, $\{Q_{\theta,n}; \theta \in \Re\}$ and $\{H_{\theta,n}; \theta \in \Re\}$ are very close to each other.

Chapter 2 introduces distances that are intended to make precise the above "close to each other." The ideas behind the possible introduction of such distances go back to Wald [1943]. They are also related to the "comparison of experiments" described by Blackwell [1951] and others. Here, following Blackwell, we shall use the name "experiment" for a family $\mathcal{E} = \{P_\theta; \theta \in \Theta\}$ of probability measures P_θ carried by a σ-field $\mathcal{A}$ of subsets of a set $\mathcal{X}$. The set Θ is often called the "parameter space." It is convenient to think of each θ as a theory that provides a stochastic model P_θ for the observation process to be carried out by the experimenter.

Note that in the preceding example of comparison between the Cauchy $\{P_{\theta,n}; \theta \in \Re\}$ and the gaussian $\{G_{\theta,n}; \theta \in \Re\}$, the parameter space is the same $\Theta = \Re$ but the Cauchy observations are in a space $\mathcal{X} = \Re^n$ while the gaussian Z_n is one dimensional. The distance introduced in Chapter 2 gives a number for any pair $\mathcal{E} = \{P_\theta; \theta \in \Theta\}$ and $\mathcal{F} = \{Q_\theta; \theta \in \Theta\}$ provided that they have the same parameter space Θ. Chapter 2 also gives, for Θ finite, a standard representation of experiments indexed by Θ in the form of their Blackwell canonical representation. It shows that, for finite fixed Θ, convergence in the sense of our distance is equivalent to convergence of the distribution of likelihood ratios.

Chapter 3 is about some technical problems that occur in the convergence of likelihood ratios. They are often simplified drastically by the use of a condition called "contiguity." That same chapter also introduces Hellinger transforms and Hellinger distances. They are particularly useful in the study of experiments where one observes many independent observations.

Chapter 4 is about gaussian experiments and Poisson experiments. By "gaussian," we mean gaussian shift experiments where the covariance structure is supposed to be known and is not part of the parameter space. The parameter space describes the possibilities available for the expectations of the gaussian distributions. The system is described in a general, infinite-dimensional setup to accommodate nonparametric or semi-parametric studies. Poisson experiments occur in many studies concerned with "point processes." They are not as mathematically tractable as the gaussian experiments, but they may be more important. Both Poisson and gaussian experiments are cases of infinitely divisible experiments, the gaussian case representing a form of degeneracy obtainable by passages to the limit. Both Poisson and gaussian experiments can be obtained in situations where one has many *independent* observations.

Some of the limit theorems available in that situation form the object of Chapter 5. It gives, in particular, a form of what Hájek and Šidák were friendly enough to call "Le Cam's three lemmas."

Chapter 6 is about the LAN conditions. LAN stands for local asymptotic normality, which really means local asymptotic approximation by a gaussian shift experiment linearly indexed by a k-dimensional space. In addition to detailing some of the consequences of the LAN conditions around a point, the chapter contains the description of a method of construction of estimates: One starts with a good auxiliary estimate θ_n^* and picks a suitable set of vectors $\{u_{n,i}; i = 0, 1, 2, \ldots, k\}$ with $u_{n,0} = 0$ and $\{u_{n,i}; i = 1, \ldots k\}$, a basis of the parameter space $\Re^k$. Then one fits a quadratic to the logarithms of likelihood ratios computed at the points $\theta_n^* + u_{n,i} + u_{n,j}$, $i, j = 0, 1, \ldots, k$. One takes for the new estimate T_n the point of $\Re^k$ that maximizes the fitted quadratic.

In the LAN case, the estimate so constructed will be asymptotically minimax, asymptotically sufficient. It will also satisfy Hájek's convolution theorem, proved here by van der Vaart's method. The chapter ends with a description of what happens in the locally asymptotically *mixed* normal (LAMN) case. Here, we cite mostly results taken from Jeganathan's papers, referring to the books by Basawa and Prakasa Rao, [1980], Basawa and Scott [1983], Prakasa Rao [1987], and Greenwood and Shiryaev [1985] for other results and examples.

Chapter 7 comes back to the case of independent observations, describing what the LAN conditions look like in that case and more particularly in the standard independent identically distributed case. Most statisticians have heard of a theory of maximum likelihood based on Cramér's conditions. The theory obtainable by application of the results of Chapter 6 is somewhat similar but more in tune with the concepts of Chapter 2. Its conditions are weaker than those used by Cramér. A sufficient condition is an assumption of differentiability in quadratic mean, discussed at some length here. The theory also works in other cases, as shown by several examples. We have also included a brief outline of Pfanzagl's theory of tangent spaces, to be used in nonparametric and semiparametric situations, together with a different formulation where parameter sets vary with the number of observations.

Chapter 8 is about Bayes procedures and the so-called Bernstein-von Mises theorem. We give a statement of a form of it with a proof intended to show the various steps involved.

Each chapter is followed by an appendix that gives a short history of the subject. We hope the history is reasonably accurate. We have attempted to give a useful list of references. The list cannot be called complete; that would be impossible in a short volume. It can be supplemented by the references contained in other books, which the interested reader should consult for additional information. Among books on related subjects, we mention the following:

1. Basawa, I. and Prakasa Rao, B.L.S. [1980] *Statistical Inference for Stochastic Processes.* Academic Press.

2. Basawa, I. and Scott, D.J. [1983] *Asymptotic Optimal Inference for Non-Ergodic Models.* Springer-Verlag.

3. Greenwood, P.E. and Shiryayev, A.N. [1985] *Contiguity and the Statistical Invariance Principle.* Gordon and Breach.

4. Ibragimov, I.A. and Has'minskii, R.Z. [1981] *Statistical Estimation. Asymptotic Theory.* Springer-Verlag.

5. Le Cam, L. [1986] *Asymptotic Methods in Statistical Decision Theory.* Springer-Verlag.

6. Pfanzagl, J. and Wefelmeyer, W. [1982] *Contributions to a General Asymptotic Statistical Theory.* Springer-Verlag.

7. Prakasa Rao, B.L.S. [1987] *Asymptotic Theory of Statistical Inference.* John Wiley & Sons.

8. Serfling, R.J. [1980] *Approximation Theorems of Mathematical Statistics.* John Wiley & Sons.

9. Strasser, H. [1985] *Mathematical Theory of Statistics.* Walter de Gruyter.

10. van de Geer, S. [2000] *Empirical Processes in M-Estimation.* Cambridge University Press.

11. van der Vaart, A. [1998] *Asymptotic Statistics.* Cambridge University Press.

2

Experiments, Deficiencies, Distances

2.1 Comparing Risk Functions

Following Blackwell [1951] we shall call experiment a family $\mathcal{E} = \{P_\theta; \theta \in \Theta\}$ of probability measures P_θ on a σ-field $\mathcal{A}$ of subsets of a set $\mathcal{X}$. The set of indices Θ is called the parameter space. To obtain a statistical decision problem in the sense of Wald we also need a set Z of possible decisions and a loss function W defined on $\Theta \times Z$ and taking values in $(-\infty, +\infty]$.

The statistician observes a value $x \in \mathcal{X}$ obtained from a probability measure P_θ. He does not know the value of θ and must select a decision $z \in Z$. He does so by choosing a probability measure ρ_x on Z and picking a point in Z at random according to ρ_x. If he has chosen z when the true distribution of x is P_θ, he suffers a loss $W_\theta(z)$. His average loss when x is observed is then $\int W_\theta(z)\rho_x(dz)$. His all over average loss when x is picked according to P_θ is the integral $\int[\int W_\theta(z)\rho_x(dz)]P_\theta(dx)$.

The function $\rho : x \rightsquigarrow \rho_x$ is called a (randomized) *decision procedure* or *decision function.* The integral

$$R(\theta, \rho) = \int [\int W_\theta(z)\rho_x(dz)]P_\theta(dx) \tag{2.1}$$

is called the *risk* of the procedure ρ for the loss function W and the true value θ. To indicate its dependence on W and on the probability measure, we shall often use the notation

$$R(\theta, \rho) = W_\theta \rho P_\theta. \tag{2.2}$$

For the above to make sense, the integrals must exist. To ensure that, we shall make the following assumptions:

(i) For each $\theta \in \Theta, \inf W_\theta(z) > -\infty$.

(ii) For each x the probability measure ρ_x is defined on a σ-field $\mathcal{D}$ in Z for which $z \rightsquigarrow W_\theta(z)$ is measurable.

(iii) For each $D \in \mathcal{D}$, the function $x \rightsquigarrow \rho_x(D)$ is $\mathcal{A}$-measurable.

A function $x \rightsquigarrow \rho_x$ that assigns for each $x \in \mathcal{X}$ a probability measure ρ_x on $(Z, \mathcal{D})$ subject to measurability requirement (iii) is often called a *Markov kernel* (from $(\mathcal{X}, \mathcal{A})$ to $(Z, \mathcal{D})$) because of the occurrence of such objects in the theory of Markov chains. Such functions also occur in conditional distributions given $\mathcal{A}$.

The σ-field $\mathcal{A}$ plays an important role. Measurability requirement (iii) reflects, in a very weak way, the intention that the procedure could in fact be carried out with the available information. It is not at all uncommon that the actual usable $\mathcal{A}$ is drastically smaller than what would be conceivable, because of poor planning, matters of privacy, lack of memory space, or other reasons.

For instance, one could conceive of recording a vector $X = (X_1, \ldots, X_n)$, but to save space or transmission time or to protect the privacy of individuals, one records only the average $\overline{X} = \sum_{j=1}^{n} X_j$. In such a case, the σ-field could have been the σ-field $\mathcal{A}_n$ of Borel subsets of the space R^n, but the available $\mathcal{A}$ is only the σ-field generated by the average function $\overline{X}$ in $\Re^n$.

A related phenomenon occurs in vital statistics, where ages, such as "age at death," are recorded only in whole numbers of years.

The structure described here, with an experiment $\mathcal{E} = \{P_\theta : \theta \in \Theta\}$ and a loss function W defined on $\Theta \times Z$, is an integral part of Wald's theory of statistical decision functions. However, we do not take into account the cost of the observations. This means that even though observations can be taken in multiple stages, we assume that continuation and stopping rules are prescribed in advance. To give an example, consider a sequence $X_1, X_2, \ldots$ of independent variables, each distributed according to a uniform distribution on $[\theta, \theta + 1]$. Fix an $\epsilon, 0 < \epsilon < 1$. Stop observing at the first n such that $\max_{j \leq n} X_j - \min_{j \leq n} X_j \geq 1 - \epsilon$. For a fixed ϵ this is an experiment in accord with our description. Different values of ϵ give different experiments. For a given loss function, these experiments will give different risk functions, but we have not added a cost for the observations. Some authors have entertained this possibility. See, for example, Espen Norberg [1999].

It is part of Wald's theory that the "quality" of a decision procedure should be judged on the risk function it induces. The choice of the loss function W, including the choice of Z in the domain $Z \times \Theta$ of W, reflects the particular statistical problem under study.

For instance, consider an experiment $\mathcal{E} = \{P_\theta : \theta \in \Theta\}$ where Θ is the real line and $P_{\theta,n}$ is the product measure, distribution of n independent random variables, $X_1, X_2, \ldots, X_n$, each with a Cauchy density $1/(\pi[1 + (x - \theta)^2])$. Take for Z the real line. Then loss functions such as $|z - \theta|$ or $|z - \theta|^2$ would correspond to estimation problems. A loss function W where $W_\theta(z) = 0$ if z and θ have the same sign (so that $z\theta \geq 0$) and $W_\theta(z) = 1$ otherwise, corresponds to a testing problem (with $\theta = 0$ considered an indifferent zone). For the testing problem, one could as well have chosen a Z consisting of two points only, say, $\{-\}$ and $\{+\}$, and let $W(-, \theta) = 1$ if $\theta > 0$ and zero otherwise. Similarly, let $W(+, \theta) = 1$ if $\theta < 0$ and zero otherwise. As this example indicates, when studying several problems at once, it may be convenient to consider that W and Z specify a certain set of functions on Θ, the set of functions $\theta \rightsquigarrow W(\theta, z); z \in Z$.

The reader may wonder why we have not restricted our attention to *nonrandomized* decision procedures, given by functions $x \rightsquigarrow z(x)$ that assign to each $x \in X$ a specific point $z(x) \in Z$ (in a measurable way). Besides the fact that one cannot prevent a statistician from tossing coins for randomization, there is a serious mathematical reason: The randomized decision procedures form a convex set. Specifically, if ϕ' and ϕ'' are two decision procedures and $\alpha \in [0, 1]$ then $\phi''' = \alpha\phi' + (1 - \alpha)\phi''$ is another decision procedure. This convexity property is essential for the derivation of many of Wald's results. See, for instance, Theorem 1 below.

Since the "quality" of decision procedures is judged by their risk functions, we are led to study more closely the space of risk functions available for an experiment $\mathcal{E}$ and a loss function W.

Let us introduce a set $\mathcal{R}(\mathcal{E}, W)$ of functions from Θ to $(-\infty, +\infty]$ obtainable in the following way: A point r belongs to $\mathcal{R}(\mathcal{E}, W)$ if there is a decision procedure ρ available on $\mathcal{E}$ such that its risk satisfies $W_\theta \rho P_\theta \leq r(\theta)$ for all θ.

To include functions that are larger than actual risk functions in the set $\mathcal{R}(\mathcal{E}, W)$ is just a matter of convenience. It allows simple wording of some results.

The set $\mathcal{R}(\mathcal{E}, W)$ has two interesting properties. The major one is a convexity property, as follows, assuming Z and Θ not empty:

(A) (*Convexity*). The set $\mathcal{R}(\mathcal{E}, W)$ is not empty. Let r_1 and r_2 be two elements of set $\mathcal{R}(\mathcal{E}, W)$, and let α be a number $\alpha \in [0, 1]$. Then there is an $r_3 \in \mathcal{R}(\mathcal{E}, W)$ such that

$$r_3(\theta) \leq \alpha r_1(\theta) + (1 - \alpha) r_2(\theta) \quad \text{for all } \theta \in \Theta.$$

The second property is used to make statements simple.

(B) If $r \in \mathcal{R}(\mathcal{E}, W)$ and g is a bounded nonnegative function, then $r + g \in \mathcal{R}(\mathcal{E}, W)$.

These properties are almost enough to obtain a minimax type theorem, but not quite. The theorem, stated as Theorem 1, involves the pointwise closure $\bar{\mathcal{R}}(\mathcal{E}, W)$ of $\mathcal{R}(\mathcal{E}, W)$ in the space of all functions from Θ to $[-\infty, +\infty]$. It also involves probability measures π with finite support on Θ. They are used as prior distributions and lead to a *Bayes risk*:

$$\chi(\pi, \mathcal{E}, W) = \inf_r \{ \int r(\theta) \pi(d\theta); r \in \mathcal{R}(\mathcal{E}, W) \}.$$

Theorem 1. *Let Θ be an arbitrary set and let $\mathcal{R}$ be a set of functions from Θ to $(-\infty, +\infty]$. Assume that $\mathcal{R}$ satisfies requirements (A) and (B) stated above. Assume also that the pointwise closure $\bar{\mathcal{R}}$ of $\mathcal{R}$ in the set of all functions from Θ to $[-\infty, +\infty]$ consists of functions with values in $(-\infty, +\infty]$.*

Let f be a function from Θ to $(-\infty, +\infty]$. Then in order for $f \in \bar{\mathcal{R}}$ it is necessary and sufficient that for every probability measure π with finite support on Θ we have

$$\int f(\theta)\pi(d\theta) \geq \chi(\pi) = \inf_r \{\int r(\theta)\pi(d\theta); r \in \mathcal{R}\}.$$

Proof. The infimum

$$\inf_r \{\int r(\theta)\pi(d\theta); r \in \bar{\mathcal{R}}\}$$

is the same number $\chi(\pi)$ that was obtained from $\mathcal{R}$. Thus the condition is certainly necessary.

In the other direction, consider *finite* subsets $F \subset \Theta$. One sees that if for each such finite set F there is an r_F that satisfies $r_F(\theta) \leq f(\theta)$ for all $\theta \in F$, then there is an $r \in \bar{\mathcal{R}}$ such that $r(\theta) \leq f(\theta)$ for all $\theta \in \Theta$.

This is because the sets $S_F = \{r; r \in \bar{R}, r_(\theta) \leq f(\theta), \text{for } \theta \in F\}$ are compact for pointwise convergence. If their intersection was empty, one of them would also be empty. Thus it is enough to look at the finite sets F and the restrictions of f and $\bar{\mathcal{R}}$ to such sets F. One can assume that f is finite on F. One proves the result directly for card $F \leq 2$. The proof can then be completed by induction on the number of elements of F. See, for instance, Kneser [1952]. □

Remark 1. Here we have allowed the elements of $\mathcal{R}$ or $\bar{\mathcal{R}}$ to take values $+\infty$. It these $+\infty$ values are not allowed, the theorem is a consequence of the Hahn-Banach theorem on separation of convex sets. Here we would need only the weak form of Hahn-Banach that says that in a locally convex vector space if C is a closed convex set and $f \notin C$ then there is a continuous linear functional ϕ such that $\phi(f) \leq \inf_g\{\phi(g); g \in C\}$, and using the first part of the preceding proof, we would need only the Hahn-Banach theorem for finite-dimensional vector spaces. The property (B) of $\mathcal{R}$ was meant to ensure that the linear functional ϕ is automatically positive. For some other formulations, see Le Cam [1964, p. 1428].

Remark 2. Theorem 1 can be applied in particular to a function $\theta \rightsquigarrow f(\theta)$ that is identically equal to a constant b. Such a function belongs to $\bar{\mathcal{R}}$ if and only if $\chi(\pi) \leq b$ for all π. This gives a *minimax* theorem

$$\sup_{\pi} \inf_{r} \int r(\theta)\pi(d\theta) = \inf_{r} \sup_{\theta} r(\theta),$$

where the $\inf_r$ is taken for $r \in \bar{\mathcal{R}}$.

There are minimax theorems valid under weaker conditions. One of them was obtained by Sion [1958]. A simple proof of Sion's theorem can be found in the book by Berge and Ghouila-Houri [1962].

The occurrence in Theorem 1 of the closure $\bar{\mathcal{R}}$ of $\mathcal{R}$ instead of $\mathcal{R}$ itself is an almost unavoidable nuisance. Thus it would be convenient if $\mathcal{R}(\mathcal{E}, W)$ would often be equal to its closure $\bar{\mathcal{R}}(\mathcal{E}, W)$. Le Cam [1964, 1986] achieves this through an extension of the definition of decision procedure, ρ_x. To see how this happens, take on Z a vector lattice Γ of bounded numerical functions. Here lattice is meant in the sense that if $\gamma_i, i = 1, 2$ are elements of Γ then their pointwise maximums $\gamma_1 \vee \gamma_2$ are also an element of Γ. Assume that Γ contains the constant functions. For an experiment $\mathcal{E} = \{P_\theta : \theta \in \Theta\}$ given by measures on $(\mathcal{X}, \mathcal{A})$, let $L = L(\mathcal{E})$ be the space of all finite signed measures that are dominated by convergent sums of the type $\sum_\theta a_\theta P_\theta, a_\theta \geq 0$. Assuming that the $\gamma \in \Gamma$ are $\mathcal{D}$ measurable, a Markov kernel $x \rightsquigarrow \rho_x$ from $(\mathcal{X}, \mathcal{A})$ to $(Z, \mathcal{D})$ induces a positive bilinear form on $\Gamma \times L$ by the recipe

$$\gamma\rho\mu = \int \int \gamma(z)\rho_x(dz)\mu(dx).$$

It is a function that is linear in γ and μ separately. It is positive in the sense that if $\rho \geq 0$ and $\mu \geq 0$, then $\gamma\rho\mu \geq 0$. It is also normalized in the sense that for $\gamma = 1$ and $\mu \geq 0$ the value $1\rho\mu$ is the total mass $\| \mu \|$ of μ.

One can read the symbol $\gamma\rho\mu$ in different ways: For $\mu \geq 0$ the function $\gamma \rightsquigarrow \gamma(\rho\mu)$ is a positive linear functional on Γ. For γ fixed, $\gamma \geq 0$, the function $\mu \rightsquigarrow (\gamma\rho)\mu$ is a positive linear functional on L. For Markov kernels, and $\mu \geq 0$, $\rho\mu$ is the measure image of μ by ρ and $\gamma\rho$ is a bounded measurable function.

Le Cam [1964] admits as (generalized) decision procedures all the positive bilinear normalized forms and call them *transitions*. It can be shown that the general transitions are, in a fairly strong sense, limits of transitions given by very simple Markov kernels (see Le Cam [1986, p. 7, theorem 1]). If one admits all *transitions* as decision procedures and defines risk appropriately (as $W_\theta \rho P_\theta = \sup\{\gamma\rho P_\theta: \gamma \in \Gamma, \gamma \leq W_\theta\}$), then the new $\mathcal{R}(\mathcal{E}, W)$ is automatically equal to $\bar{\mathcal{R}}(\mathcal{E}, W)$. This is because *transitions* are defined by properties of "finite character" that are preserved by taking limits.

The restrictions imposed in the classical books on decision theory imply that one can choose Γ so that all bounded positive linear functionals on it are given by measures, that all bounded positive linear functionals on L are given by bounded measurable functions, and that one can "lift" them. That is, one can choose in the equivalence class of such a function a special representative function in such a way that positivity and linearity requirements are preserved. Then all our transitions are representable by Markov kernels. To ensure that Theorem 1 applies to $\mathcal{R}(\mathcal{E}, W)$ itself, without closure, one needs to impose a further restriction.

A possibility is to require the following:

(C) Let ϕ be a function from Θ to $(-\infty, +\infty]$. Suppose that for every finite set $F \subset \Theta$ there is a $z_F \in Z$ such that $W_\theta(z_F) \leq \phi(\theta)$ for all $\theta \in F$. Then there is a $z \in Z$ such that $W_\theta(z) \leq \phi(\theta)$ for all $\theta \in \Theta$.

The condition that all transitions be representable by Markov kernels is awkward. It does not eliminate certain "nonparametric" problems, only most of them. For instance, in those nonparametric problems that involve "all" distributions on the line, the representation of transitions by Markov kernels is often impossible. If, on the contrary, you take "all distributions that have a Lebesgue density," you have a chance. For general results, it appears to be easier to work with transitions, even if in the end one has to check that the particular transition selected for a special purpose can indeed be represented by a Markov kernel.

With these preliminaries, we are ready to define a "distance" between experiments that corresponds to the same parameter set Θ.

2.2 Deficiency and Distance between Experiments

Previously we allowed our risk functions and even the loss function W to take arbitrary values in $(-\infty, +\infty]$. This will be useful in a later chapter, for instance, when we deal with the so-called asymptotic minimax theorem.

Now we shall allow only loss functions W such that $0 \leq W_\theta(z) \leq 1$. This is because multiplying the loss function by a constant $c \geq 0$ would multiply our "distance" by the same constant c. It could then be made as large as one pleases unless the "distance" is zero.

Consider the situation of an experimenter who could carry out an experiment $\mathcal{E} = \{P_\theta : \theta \in \Theta\}$ or an experiment $\mathcal{F} = \{Q_\theta : \theta \in \Theta\}$ but not both. The set Θ represents the set of theories held by the experimenter about the "states of the nature". It is not supposed to be affected by the choice of experiment to be conducted. However $\mathcal{E}$ involves a set $\mathcal{X}$ with σ-field $\mathcal{A}$. The experiment $\mathcal{F}$ may involve a totally different set $\mathcal{Y}$ with a σ-field $\mathcal{B}$. For instance $\mathcal{E}$ and $\mathcal{F}$ might be about the life of light bulbs. In $\mathcal{E}$, one takes m light bulbs and counts how many fail in a year. In $\mathcal{F}$ one takes n light bulbs and records the times at which the first 10 failures occur. The choice between $\mathcal{E}$ and $\mathcal{F}$ may depend on what the experimenter knows about the life of light bulbs, but it does not, by itself, affect theories about light bulb failures.

Definition 1. *(Deficiency). The deficiency $\delta(\mathcal{E}, \mathcal{F})$ of $\mathcal{E}$ with respect to $\mathcal{F}$ is the smallest number $\delta \in [0, 1]$ such that for every arbitrary loss function W with $0 \leq W \leq 1$ and every $r_2 \in \mathcal{R}(\mathcal{F}, W)$ there is an $r_1 \in \bar{\mathcal{R}}(\mathcal{E}, W)$ such that $r_1(\theta) \leq r_2(\theta) + \delta$*

for all $\theta \in \Theta$.

This can be stated differently: For every W, $0 \leq W \leq 1$ the set

$$\bar{\mathcal{R}}(\mathcal{E}, W) \quad \text{contains} \quad \{\mathcal{R}(\mathcal{F}, W) + \delta\}.$$

(A symbol such as $\mathcal{R} + \delta$ means the set of functions $\{r + \delta; r \in \mathcal{R}\}$.) Note that we have said "smallest number." It is an easy consequence of the compactness of $\bar{\mathcal{R}}(\mathcal{E}, W)$ that there exists such a smallest number. Alternatively, we could have defined $\delta(\mathcal{E}, \mathcal{F})$ as the infimum of the set of numbers $\delta \in [0, 1]$ such that $\mathcal{R}(\mathcal{F}, W) + \delta \subset \bar{\mathcal{R}}(\mathcal{E}, W)$. (For symmetry, one can replace $\mathcal{R}(\mathcal{F}, W)$ by $\bar{\mathcal{R}}(\mathcal{F}, W)$.)

Definition 2. *The "distance"* $\Delta(\mathcal{E}, \mathcal{F})$ *between the two experiments* $\mathcal{E}$ *and* $\mathcal{F}$ *is the maximum of* $\delta(\mathcal{E}, \mathcal{F})$ *and* $\delta(\mathcal{F}, \mathcal{E})$.

This "distance" is not actually a metric but only a pseudometric. It satisfies the triangle inequality $\Delta(\mathcal{E}, \mathcal{G}) \leq \Delta(\mathcal{E}, \mathcal{F}) + \Delta(\mathcal{F}, \mathcal{G})$ but the equality $\Delta(\mathcal{E}, \mathcal{F}) = 0$ does not imply that $\mathcal{E}$ and $\mathcal{F}$ are actually the same. If $\Delta(\mathcal{E}, \mathcal{F}) = 0$ we shall call $\mathcal{E}$ and $\mathcal{F}$ *equivalent* or of the *same type*. The class of experiments corresponding to a given Θ does not constitute a set in the usual axioms of set theory. However, their *types* constitute a set, and Δ is a metric on the set of experiment types indexed by Θ.

Note the wording "every W" subject to $0 \leq W \leq 1$. This means a lot of possible statistical problems, even if one considers only those problems in which the set Z of possible decisions is a compact set on which W is continuous, or even if one uses only sets Z that are finite. To restrict the sets Z to be finite would not change the value of $\delta(\mathcal{E}, \mathcal{F})$ because we use the closure $\bar{\mathcal{R}}(\mathcal{E}, W)$ instead of $\mathcal{R}(\mathcal{E}, W)$, and because we would not impose any limitation on the cardinality of the finite sets Z. For distances Δ_k defined in a similar manner but restricting the sets Z to have cardinality at most equal to a specified integer k, see, Torgersen [1991].

Lehmann [1988] has argued, with some justice, that to allow "every W" is to allow too many statistical problems. However, we do not know of any natural restrictions to place on the "every W" and do not know what an experimenter might want to contemplate. So we shall take the above definitions of deficiency and distance.

To summarize the intent of the definitions, if you allow only loss functions W such that $0 \leq W \leq 1$ and $\Delta(\mathcal{E}, \mathcal{F}) = \epsilon$, any risk function you can have on one of the two experiments ($\mathcal{E}$ or $\mathcal{F}$) can be matched within ϵ by a risk function on the other experiment.

It turns out that our deficiency can also be obtained in a different way that does not explicitly involve loss functions. Define a norm for signed measures μ by

$$\|\mu\| = \sup_f \{ \int f d\mu; f \quad \text{measurable and} \quad |f| \leq 1 \}$$

(often called the L_1-norm).

Consider transitions K from measures on $\mathcal{A}$ to measures on $\mathcal{B}$. If the transition K is given by a Markov kernel, KP, the image KP evaluated at $B \in \mathcal{B}$ is defined by $KP_\theta(B) = \int K(x, B)$ $P_\theta(dx)$ for $x \in \mathcal{A}$. If K is a general transition, KP is a limit $\lim_\nu K_\nu P$ for Markov kernels K_ν. Then we have the following theorem.

Theorem 2. *The deficiency $\delta(\mathcal{E}, \mathcal{F})$ is exactly*

$$\delta(\mathcal{E}, \mathcal{F}) = \inf_K \sup_\theta \frac{1}{2} \|Q_\theta - KP_\theta\|,$$

where the infimum is taken over the set of all transitions K from measures on $\mathcal{A}$ to measures on $\mathcal{B}$ (or from the spaces $L(\mathcal{E})$ to $L(\mathcal{F})$, where $L(\mathcal{E})$ and $L(\mathcal{F})$ are defined after Remark 2).

The proof is somewhat complicated and will not be given here. For the proof, see Le Cam [1986, p. 20-23].

This theorem extends a previous result of Blackwell, Sherman and Stein [1951] related to the case where $\delta(\mathcal{E}, \mathcal{F}) = 0$. See also Strassen [1965].

If one takes into account the fact that our transitions K are slight extensions of randomizations through Markov kernels,

Theorem 2 says that $\delta(\mathcal{E}, \mathcal{F}) = \epsilon$ only if one can (almost) reproduce the measures Q_θ of $\mathcal{F}$ within 2ϵ through randomizations K independent of Θ.

The definition of a distance between experiments, just given, obviously has a very specific statistical meaning. Even the alternate definition implied by Theorem 2 has such a meaning. If Q and KP are two probability measures on the same σ-field $\mathcal{B}$ then $\|Q - KP\| = 1 - (1/2)\|Q - KP\|$ is the minimum sum of probabilities of errors for tests between Q and KP.

Thus if $\|Q - KP\|$ is very small, say, $\|Q - KP\| \leq 10^{-10^{137}}$, there is no earthly way to distinguish between Q and KP for all practical purposes.

The case where $\delta(\mathcal{E}, \mathcal{F}) = 0$ has received a lot of attention, leading to the following terminology. If $\delta(\mathcal{E}, \mathcal{F}) = 0$ one says that $\mathcal{F}$ is less informative than $\mathcal{E}$, or that $\mathcal{E}$ is better than $\mathcal{F}$, or that $\mathcal{E}$ is more informative than $\mathcal{F}$.

The reader should beware of the fact that even if our deficiency δ has a perfectly reasonable statistical meaning, it is not readily computable in most instances. All one can hope in most cases is to find upper and lower bounds for it. Still, it can be useful. To see what happens consider a population of N individuals, with N large. The proportion of people in the population that would say that they will vote for candidate N. G. in the next presidential election is a certain number p. To ascertain p, the investigator takes a sample of 1000 individuals. This can be done at random without replacement, yielding our experiment $\mathcal{E}_N$. It can also be done at random *with* replacement, yielding another experiment $\mathcal{F}_N$. It stands to reason that $\delta(\mathcal{E}_N, \mathcal{F}_N) = 0$, and indeed that can be proved. On the contrary the value of $\delta(\mathcal{F}_N, \mathcal{E}_N)$ is harder to ascertain. However, one can readily argue that it cannot be much higher than the probability of having duplicates in the sample with replacement. One can also argue that $\delta(\mathcal{F}_N, \mathcal{E}_N)$ tends to zero as $N \to \infty$.

In spite of the inherent difficulties in computing $\delta(\mathcal{E}, \mathcal{F})$, Torgersen [1972] was able to give computable expressions for it for experiments satisfying some invariance properties. Hansen and Torgersen [1974] did the analysis for linear models. In one of the cases considered by Torgersen [1974], one observes $n + r$ i.i.d. variables, each distributed as $\mathcal{N}(\theta, \mathrm{I})$, with I the identity matrix for k dimensions. This gives our experiment $\mathcal{E}$. Another observer sees only the values of the first n variables, resulting in an experiment $\mathcal{F}$.

Then, according to Torgersen,

$$\Delta(\mathcal{E}, \mathcal{F}) = P\{\log \frac{1+\alpha}{\alpha} \leq \frac{\chi^2}{k} \leq \frac{1+\alpha}{\alpha} \log(1+\alpha)\},$$

where $\alpha = r/n$ and χ^2 has the central χ^2 distribution with k degrees of freedom. For α small this is roughly proportional to $(r/n)\sqrt{k}$, indicating that for r/n small and k moderate a few additional observations do not bring much additional information.

We shall have occasion in Chapter 6 to consider another number called *insufficiency*. It behaves more like $[kr/n]^{1/2}$ and measures how far the first n observations are to being "sufficient" for the $n + r$ ones. That is a quite a different question. Our *deficiency* does not in any way presume that our experiment $\mathcal{E}$ is a subexperiment of $\mathcal{F}$.

Another example where exact computations are difficult to carry out but bounds can be obtained is as follows.

Let $\mathcal{E}_n = \{P_{\theta,n} : \theta \in \Theta\}$ where $P_{\theta,n}$ is a product measure, distribution of n i.i.d. variables with a Cauchy density $1/(\pi[1 + (x-\theta)^2])$ on the line. Here Θ is taken to be the real line $\Theta = (-\infty, +\infty)$. From $\mathcal{E}_n$ one can construct another experiment $\mathcal{F}_n$ where one retains from $\mathcal{E}_n$ only the value $\hat{\theta}_n$ of a maximum likelihood estimate. One can also consider another experiment $\mathcal{G}_n$ where one observes n i.i.d. variables that are individually Gaussian, $\mathcal{N}(\theta, 2)$. It will be a consequence of the arguments of Chapter 5 that $\Delta(\mathcal{E}_n, \mathcal{F}_n)$ and $\Delta(\mathcal{E}_n, \mathcal{G}_n)$ tend to zero as $n \to \infty$ and this is at a speed of order roughly $1/\sqrt{n}$. Thus, one can study the possible risks in $\mathcal{E}_n$ to a certain approximation by looking at what happens in $\mathcal{G}_n$,

A recent result of M. Nussbaum involves a different nonparametric situation. Nussbaum considers the set Θ of densities with respect to Lebesgue measure on $[0, 1]$ that satisfy a

Hölder condition $|\theta(x) - \theta(y)| \leq M|x - y|^\alpha, \alpha > 1/2$ and a bound $\theta(x) \geq \epsilon > 0$. This gives, for n i.i.d. observations our experiment $\mathcal{E}_n$.

Let $\mathcal{G}_n$ be the gaussian experiment where one observes a process $\{Y(t); t \in [0, 1]\}$ such that $dY(t) = \sqrt{\theta(t)} + (1/2\sqrt{n})dW(t)$ where W is the standard Wiener process on $[0, 1)$. Nussbaum shows that $\Delta(\mathcal{E}_n, \mathcal{G}_n) \to 0$ and gives bounds, lower and upper, on $\Delta(\mathcal{E}_n, \mathcal{G}_n)$.

To terminate this section, let us mention some immediate corollaries of the definitions and Theorem 1.

Take $\mathcal{E}$ and $\mathcal{F}$, arbitrary, with the same parameter set Θ.

Corollary 1. *The deficiency $\delta(\mathcal{E}, \mathcal{F})$ is the smallest number $\epsilon \in [0, 1]$ such that for every loss function W, $0 \leq W \leq 1$, one has*

$$\bar{\mathcal{R}}(\mathcal{F}, W) + \epsilon \subset \bar{\mathcal{R}}(\mathcal{E}, W).$$

Corollary 2. *The deficiency $\delta(\mathcal{E}, \mathcal{F})$ is the smallest number $\epsilon \in [0, 1]$ such that the Bayes risks satisfy $\chi(\pi, W, \mathcal{E}) \leq \chi(\pi, \mathcal{W}, \mathcal{F}) + \epsilon$ for all W, $0 \leq W \leq 1$ and all probability measures π with finite support on Θ.*

The distance $\Delta(\mathcal{E}, \mathcal{F})$ is equal to

$$\sup_{\pi, W} |\chi(\pi, W, \mathcal{E}) - \chi(\pi, \mathcal{W}, \mathcal{F})|$$

where the supremum is taken over all $W, 0 \leq W \leq 1$ and all probability measures π with finite support on Θ.

Now we move on to something that appeals to almost every statistician and show that our distance is related to something involving the distribution of likelihood ratios.

2.3 Likelihood Ratios and Blackwell's Representation

In this section, unless the contrary is explicitly stated, it will be assumed that Θ is a finite set containing k elements.

Consider an experiment $\mathcal{E} = \{P_\theta : \theta \in \Theta\}$, where the P_θ are probability measures on a space $(\mathcal{X}, \mathcal{A})$. Let $S = \sum_\theta P_\theta$; $\theta \in \Theta$ be the sum of the measures P_θ. Since S dominates each P_θ, one can form the Radon-Nikodym densities $f_\theta = (dP_\theta/dS)$. Evaluated at $x \in \mathcal{X}$ these densities yield a vector $v(x) = \{f_\theta; \theta \in \Theta\}$ of likelihood ratios with values in $\Re^k$. One can assume that $f_\theta(x) \geq 0$ and that $\sum_\theta f_\theta(x) \equiv 1$, for all $x \in \Re^k$. Thus $v(x)$ is a point in the unit simplex $U(\Theta)$, the set of vectors $u = \{u_\theta : \theta \in \Theta\}$ such that $u_\theta \geq 0$, and $\sum_\theta u_\theta = 1$.

The image $m_\mathcal{E}$ on $U(\Theta)$ of the measure S by the map $x \rightsquigarrow v(x)$ is called the Blackwell canonical measure of $\mathcal{E}$. (In fact Blackwell [1951] used S/k, but this is immaterial).

In Theorem 2 of Section 2.2, we saw that the risks obtainable from an experiment $\mathcal{E}$ can be characterized through the Bayes risks $\chi(\pi, \mathcal{E}, W)$. Our first task here will be to describe these Bayes risks. This was done by Torgersen [1970]. Our description derives from his but is perhaps a bit simpler. The results are given in Lemmas 1 and 2; they lead to the main result of this section, Theorem 3, which asserts that the distance Δ introduced in the previous section is equivalent to the dual Lipschitz distance between Blackwell's canonical measures.

First it is convenient to work with *gains* instead of *losses*. Thus, instead of a loss function W, $0 \leq W \leq 1$ we shall use a gain function $V = 1 - W$. (This gives a simpler description of the class Φ_1 introduced later.)

To get a Bayes gain one would maximize the expression with respect to ρ

$$\int \pi(d\theta) \int \int V_\theta(z) \rho_x(dz) P_\theta(dx).$$

Assuming that π gives mass π_θ to θ and interchanging finite sums and integrals, the quantity to maximize becomes

$$\int S(dx) \int \rho_x(dz) [\sum_\theta \pi_\theta V_\theta(z) f_\theta(x)] \tag{2.3}$$

for the densities $f_\theta = dP_\theta/dS$.

Clearly, this will be maximized if for each x one can find a $z(x)$ such that

$$\sup_z \sum_\theta \pi_\theta V_\theta(z) f_\theta(x) = \sum_\theta \pi_\theta V_\theta(z(x)) f_\theta(x).$$

Now note that the quantity in square brackets in (2.3) depends on x only through the vector of densities $v(x) = \{f_\theta(x); \theta \in \Theta\}$. Thus, one could instead maximize for each $u \in U(\Theta)$ the sums $\sum_\theta \pi_\theta V_\theta(z) u_\theta$. Letting

$$\phi(u) = \sup_z \sum_\theta \pi_\theta V_\theta(z) u_\theta, \tag{2.4}$$

one would then substitute for u the vector $v(x)$. As one varies the prior measure π and the gain function V, the functions $\phi(u)$ form a certain class that will be denoted by Φ_1.

As we shall see later, it turns out that these functions ϕ are measurable; they even satisfy Lipschitz conditions, which has further consequences. If no $z \in Z$ that achieves the maximum, one can still get as close as one wants to the supremum given by ϕ. Thus our Bayes gain for π and V would be

$$\sup_\rho \int \pi(d\theta) \int \int V_\theta(z) \rho_x(dz) P_\theta(dx) = \int \phi[v(x)] S(dx).$$

This obviously depends only on the Blackwell measure $m_\mathcal{E}$ and can be written as

$$\int \phi[v(x)] S(dx) = \int \phi(u) m_\mathcal{E}(du). \tag{2.5}$$

A function such as $\phi(u)$ can be written in the form

$$\phi(u) = \sup_z \sum_\theta c(\theta, z) u_\theta,$$

where the $c(\theta, z)$ are coefficients restricted to satisfying $0 \leq c(\theta, z)$ and $\sum_\theta \sup_z c(\theta, z) \leq 1$ (because the gain satisfies $0 \leq V_\theta(z) \leq 1$). Conversely any such function $\sup_z \sum_\theta c(\theta, z) u_\theta$ with coefficients satisfying these restrictions correspond to a statistical problem with a certain prior π and gain function V. Indeed, let $s = \sum_\theta \sup_z c(\theta, z)$. If $s = 0$, the situation is trivial, with $V \equiv 0$. If $s > 0$, define π_θ by $\sup_z c(\theta, z)/s$ and then define V by

$$\pi_\theta V_\theta(z) = c(\theta, z)$$

for those π_θ that are not zero.

To give some properties of the class Φ_1 of such functions, use on $U(\Theta)$ the distance $|u - u'| = \sup_\theta\{|u_\theta - u'_\theta|$. The properties are needed for the derivation of Lemma 3 and Theorem 3.

Lemma 1. *The class Φ_1 is a convex set of convex functions on $U(\Theta)$. If $\phi_i \in \Phi_1, i = 1, 2$, then so does $(1/2)(\phi_1 \vee \phi_2)$ where $\phi_1 \vee \phi_2$ is the pointwise maximum of ϕ_1 and ϕ_2. All the $\phi \in \Phi_1$ satisfy the Lipschitz condition*

$$|\phi(u) - \phi(u')| \leq |u - u'|.$$

Proof. Indeed, the linear functions $\sum_\theta c(\theta, z)u_\theta$ with $c(\theta, z) \geq 0$ and $\sum_\theta \sup_z c(\theta, z) \leq 1$ satisfy the stated Lipschitz condition. That condition is preserved by taking pointwise suprema. If ϕ_i corresponds to coefficients $c_i(\theta, z), z \in Z_i, i = 1, 2$, one can make other coefficients $c_3(\theta, z)$ with $z \in Z_1 \cup Z_2$ and $c_3(\theta, z) = c_1(\theta, z)/2$ if $z \in Z_1$ and $c_3(\theta, z) = c_2(\theta, z)/2$ if $z \in Z_2$. Then $\phi_3 = (1/2)(\phi_1 \vee \phi_2)$ is defined by

$$\phi_3(u) = \sup_z \sum_\theta c(\theta, z)u_\theta,$$

and similarly for the convexity of Φ_1. For $\alpha \in [0, 1]$ let $c_3(\theta, z) = \alpha c_1(\theta, z) + (1 - \alpha)c_2(\theta, z)$, where now $z = (z_1, z_2) \in Z_1 \times Z_2$. The result follows.

The fact that the $\phi \in \Phi_1$ satisfies the Lipschitz condition $|\phi(u) - \phi(u')| \leq |u - u'|$ makes them measurable, even if the $\sup_z$ is taken over an infinite set. Thus $x \rightsquigarrow \phi[v(x)]$ is always measurable, as was stated previously.

We already noted in (2.5) that for an experiment $\mathcal{E} = \{P_\theta; \theta \in \Theta\}$, Θ finite, the Bayes gains take the form

$$\int \phi(u) m_\mathcal{E}(du)$$

where $m_{\mathcal{E}}$ is the Blackwell measure of $\mathcal{E}$. Any positive measure m on $U(\Theta)$ such that $\int u_\theta m(du) = 1$ for the coordinate functions u_θ defines an experiment by taking the measures $Q_\theta(du) = u_\theta m(du)$. In particular $m_{\mathcal{E}}$ defines such an experiment $\mathcal{F} = \{Q_\theta; \theta \in \Theta\}$ for $Q_\theta(du) = u_\theta m_{\mathcal{E}}(du)$. This leads to the following observation.

Lemma 2. *The distance between two experiments* $\mathcal{E} = \{P_\theta; \theta \in \Theta\}$ *and* $\mathcal{G} = \{R_\theta; \theta \in \Theta\}$ *is*

$$\Delta(\mathcal{E}, \mathcal{G}) = \sup_\phi | \int \phi(u)[m_{\mathcal{E}}(du) - m_{\mathcal{G}}(du)]|; \phi \in \Phi_1.$$

In particular two experiments with the same Blackwell measure are equivalent. Indeed they have the same Bayes gains or risks sets $\mathcal{R}(\mathcal{E}, W) = \mathcal{R}(\mathcal{G}, W)$ *for any positive, bounded* W.

This applies in particular to the experiment $\mathcal{E}$ *and its canonical representation* $\mathcal{F} = \{Q_\theta; \theta \in \Theta\}$ *with* $Q_\theta(du) = u_\theta m_{\mathcal{E}}(du)$.

That last statement could have been seen more directly as follows. First $\Delta(\mathcal{E}, \mathcal{F}) = 0$, because anything that can be done knowing $v(x)$ can also be done knowing x. Second, the vector function $x \rightsquigarrow v(x)$ is a sufficient statistic in the Halmos and Savage [1949] sense for the family $\{P_\theta; \theta \in \Theta\}$ on $(\mathcal{X}, \mathcal{A})$. It is an accepted "principle" that sufficient statistics contain all the "information" needed. This could be argued in our framework by replacing Markov kernels $x \rightsquigarrow \rho_x$ by their conditional expectation given the sufficient statistic v, that is, by replacing $\rho_x(B)$ by $E[\rho_x(B)|v(x)]$. Unfortunately, if one wants to stick to the Markov kernel formulation, this leads to some complications: The conditional expectation of $x \rightsquigarrow \rho_x$ might not be representable by a Markov kernel. Such things happen if the sets Z are badly constituted. For instance, one might attempt to reproduce the P_θ on $(\mathcal{X}, \mathcal{A})$ from a randomization given the sufficient statistic $x \rightsquigarrow v(x)$. This might be impossible in some cases. For an example take measures P'_θ on $[0, 1]$, all dominated by the Lebesgue measure.

Take for $\mathcal{X}$ a set of inner Lebesgue measure zero and outer Lebesgue measure unity. Take for $\mathcal{A}$ the intersections of the

Borel sets of $[0, 1]$ with $\mathcal{X}$ and let P_θ be the (σ–additive!) measure defined by the outer measure of $A \in \mathcal{A}$ for P'_θ.

Such difficulties are, of course, of no importance practically speaking. However they do not occur for the "transitions" of Section 2.1. That is one of the reasons we introduced these transitions and why we did not use the categories of Morse and Sacksteder [1966], where the morphisms are Markov kernels.

This being as it may, let us move on to some more interesting inequalities satisfied by the distance Δ of Section 2.2.

We have used on $U(\Theta)$ the distance $|u - u'| = \sup_\theta |u_\theta - u'_\theta|$. As is true for any metric space, such a distance can be used to define a distance between *measures* carried by $U(\Theta)$. Here we shall concentrate on the so-called dual-Lipschitz norm. If μ is a finite signed measure carried by $U(\Theta)$, its dual-Lipschitz norm is

$$\|\mu\|_D = \sup_f |\int f d\mu|,$$

where the supremum is taken over all functions f satisfying $|f| \leq 1$ and $|f(u) - f(u')| \leq |u - u'|$. The norm defines a distance between μ_1 and μ_2 equal to $\|\mu_1 - \mu_2\|_D$. (The reader should not confuse this norm with the L_1-norm $\|\mu\| = \sup_f |\int f d\mu|$ where f is only restricted to be measurable and such that $|f| \leq 1$. Of course $\|\mu\|_D \leq \|\mu\|$ but they may have very different values.)

Take then two experiments $\mathcal{E}$ and $\mathcal{F}$ and their Blackwell measures $m_\mathcal{E}$ and $m_\mathcal{F}$. Since by Lemma 1 all the functions in Φ_1 satisfy the Lipschitz condition $|\phi(u) - \phi(u')| \leq |u - u'|$, it follows from Lemma 2 that

$$\Delta(\mathcal{E}, \mathcal{F}) \leq \|m_\mathcal{E} - m_\mathcal{F}\|_D.$$

To link $\Delta(\mathcal{E}, \mathcal{F})$ and $\|m_\mathcal{E} - m_\mathcal{F}\|_D$ more specifically, we shall need two lemmas of separate interest.

Lemma 3. *A Blackwell canonical measure is uniquely defined by its values on* Φ_1.

Proof. Let Φ be the cone of functions of the type $a\phi, a \geq 0, \phi \in \Phi_1$, and let $\mathcal{H}$ be the vector space of differences $\phi_1 - \phi_2, \phi_i \in \Phi, i = 1, 2$.

We have seen in Lemma 1 that Φ is convex and $\phi_i \in \Phi$ implies $\phi_1 \vee \phi_2 \in \Phi$. This implies that $\mathcal{H}$ is a vector lattice, since, for $\phi_i \in \Phi$,

$$[\phi_1 - \phi_2]^+ = (\phi_1 \vee \phi_2) - \phi_2.$$

A Blackwell measure m defines a positive linear functional μ on the set Φ_1, hence on the cone Φ and therefore on $\mathcal{H}$ by linearity. That functional is σ-smooth. That is, if $\{\phi_n\}$ is a sequence, $\phi_n \in \mathcal{H}$, such that ϕ_n decreases to zero pointwise on $U(\Theta)$, then $\mu(\phi_n)$ tends to zero. (It has become traditional to use the term "σ-smooth " for functions of functions, reserving the corresponding term "σ-additive" for functions of sets.) Here, the functions in $\mathcal{H}$ are Lipschitzian, hence at least continuous. If a sequence ϕ_n decreases to zero pointwise, it does so uniformly, by Dini's theorem. Hence we have the desired smoothness, since $\|\mu\|$ is bounded. One can therefore extend μ to its space of integrable functions by the procedure due to Daniell [1917, 1918]. Here the constant functions belong to $\mathcal{H}$, because $\sum_\theta u_\theta = 1$. Thus all the bounded functions that are measurable with respect to the σ-field generated by the elements of $\mathcal{H}$ are integrable in the sense of Daniell. The Daniell extension is uniquely defined by the properties of being linear, positive, and σ-smooth, hence the uniqueness results.

Note. For the convenience of the reader, let us recall the steps involved in the Daniell procedure. Starting with a vector lattice (for pointwise operations) such as $\mathcal{H}$, one defines an upper class of functions. A g belongs to the upper class if it is the pointwise supremum of a countable subset of $\mathcal{H}$. Changing the sign, one obtains the lower class. One defines $\bar{\mu}(g)$ for g in the upper class as the supremum $\sup\{\mu(h); h \in \mathcal{H}, h \leq g\}$. For our f in the lower class one writes $\underline{\mu}(f) = -\bar{\mu}(g)$ with $g = -f$. Then one declares "integrable" any function β such that for every $\epsilon > 0$ there is a lower function f and an upper g for which $f \leq \beta \leq g$ and $\bar{\mu}(g) - \underline{\mu}(f) < \epsilon$, $\bar{\mu}(g)$ being finite. The extension of μ to β is then the common point of the brackets $[\underline{\mu}(f), \bar{\mu}(g)]$. The fact that the extension is linear, positive, and σ-smooth is not immediate. For a proof, see Loomis [1953] or Le Cam [1986, p.

687].

Remark 3. The Blackwell measures are σ-additive whether the P_θ are or not. Le Cam [1986] does not require the σ-additivity of the constituent measures P_θ. They may, for instance, be cylindrical measures. Such objects crop up when the observations are stochastic processes. Blackwell measures are then defined directly from the values of the Bayes gains, but one has then to check that they are indeed well defined. For this, see Le Cam [1986, p. 24–26]. They are any way automatically σ-smooth on $\mathcal{H}$, as implied by Dini's theorem.

One could build a proof without using the Daniell procedure. It is done by using the fact that the unit ball of $\mathcal{H}$ is dense in the corresponding unit ball of the space of continuous functions on the compact $U(\theta)$, as follows from the Stone-Weierstrass theorem.

Lemma 4. *The Blackwell measures form a compact set for the metric* $\|m_\mathcal{E} - m_\mathcal{F}\|_D$.

Proof. This is a well-known fact, a consequence of more general results. (See, for instance, Dudley [1978, 1989]). A proof can be constructed along the following steps. Take an $\epsilon > 0$ and an $M, 0 < M < \infty$. Let Λ_M be the set of functions f defined on $U(\Theta)$, such that $|f| \leq M$ and $|f(u) - f(u')| \leq M|u - u'|$ for all pairs in $U(\Theta)$. Take a set of points $F_M = \{x_1, \ldots, x_M\}$ in $U(\Theta)$ such that any $u \in U(\Theta)$ is a distance at most ϵ/M^2 of one of the $x'_j s$. If a function of Λ_M is known on F_M, it is also known within ϵ/M on all of $U(\Theta)$. Thus one can approximate Λ_M within ϵ/M by a finite set, say, $\mathcal{H}_M$, of functions of Λ_M. One can assume that the coordinate functions $u_\theta, \theta \in \Theta$ are $\mathcal{H}_M$. Any Blackwell measure μ can be evaluated at the elements of $\mathcal{H}_M$. These evaluations yield, as μ varies, a certain bounded subset of a Euclidean space. Thus, given a sequence $\{\mu_n\}$ of Blackwell measures one can extract a subsequence $\{\mu_\nu\}$ such that $\int f d\mu_\nu$ converges to some limit $a(f)$ for every $f \in \mathcal{H}_M$.

This can be done for any value of M; by a diagonalization procedure; one can select a subsequence $\{\mu_m\}$ such that $\int f d\mu_m$

converges to a limit $a(f)$ for any $f \in \mathcal{H}_M$ and every M. Since $\cup_M \mathcal{H}_M$ is dense for uniform convergence in the space Λ of Lipschitz functions, $\int f d\mu_m$ converges to a limit $a(f)$, for every $f \in \Lambda$. The limit $a(f)$, being limit of positive linear functionals, is also positive and linear. Since we have retained the coordinate functions u_θ in each $\mathcal{H}_M$, the limit satisfies $a(u_\theta) = 1$. By Dini's theorem it is σ-smooth on Λ and therefore can be written in the form $a(f) = \int f d\mu$, where μ is a Blackwell measure on $U(\Theta)$.

It remains to be proved that the convergence is uniform on Λ_1. It is certainly uniform on the *finite* set $\mathcal{H}_1$. Thus

$$\limsup_m \|\mu_m - \mu\|_D \leq \epsilon.$$

Moreover ϵ can be chosen arbitrarily, hence the result.

Note. The reader who is acquainted with "ultra filters" or "universal nets" can simplify this proof by taking immediately an ultra filter $\{\mu_\nu\}$ that converges for all $f \in \Lambda$.

These two lemmas yield the main result of this section, which is as follows.

Theorem 3. *Let Θ be finite, of cardinality k. Then there exists a numerical function ω_k defined on $[0,1]$ such that $\omega_k(t) \to 0$ as $t \to 0$ and such that*

$$\Delta(\mathcal{E}, \mathcal{F}) \leq \|m_\mathcal{E} - m_\mathcal{F}\|_D \leq \omega_k[\Delta(\mathcal{E}, \mathcal{F})]$$

for all pairs of experiments indexed by Θ.

Proof. We have already seen that

$$\Delta(\mathcal{E}, \mathcal{F}) = \sup_{\phi \in \Phi_1} |\int \phi dm_\mathcal{E} - \int \phi m_\mathcal{F}| \leq \|m_\mathcal{E} - m_\mathcal{F}\|_D.$$

For the reverse inequalities, define

$$\omega_k(t) = \sup\{\|m_\mathcal{E} - m_\mathcal{F}\|_D, \Delta(\mathcal{E}, \mathcal{F}) \leq t\},$$

where the sup is over pairs $(\mathcal{E}, \mathcal{F})$.

Suppose that $\omega_k(t)$ does not decrease to zero as t decreases to zero. Then there is some ϵ such that $\omega_k(t) \geq 2\epsilon$ for $t > 0$.

Therefore these are pairs $(\mathcal{E}_n, \mathcal{F}_n)$ such that $\Delta(\mathcal{E}_n, \mathcal{F}_n) \leq 1/n$ but $\|m_{\mathcal{E}_n} - m_{\mathcal{F}_n}\|_D \geq \epsilon$.

By the compactness of Lemma 4 one can assume that $\|m_{\mathcal{E}_n} - m_1\|_D$ and $\|m_{\mathcal{F}_n} - m_2\|_D$ tend to zero for some Blackwell measures $m_i, i = 1, 2$. One must then have $\|m_1 - m_2\|_D \geq \epsilon$ from the assumed difference between $m_{\mathcal{E}_n}$ and $m_{\mathcal{F}_n}$. However, $\Delta(\mathcal{E}_n, \mathcal{F}_n) \leq 1/n$ implies that m_1 and m_2 correspond to the same experiment. Thus, by Lemma 3, $m_1 = m_2$. This contradiction proves that $\lim_{t\to 0} \omega_k(t) = 0$. □

We have indexed ω by the cardinality k of Θ, because by permuting the $\theta \in \Theta$ around, it is clear that ω depends only on k. Unfortunately, we have very little information on how ω_k depends on k. Torgersen [1991] has given various explicit expressions of $\Delta(\mathcal{E}, \mathcal{F})$ when card $\Theta = 2$. For $k > 2$, the expressions are not explicit enough to yield our ω_k.

Corollary 3. *For Θ fixed and finite the experiment types indexed by Θ form a compact set for the distance Δ.*

2.4 Further Remarks on the Convergence of Distributions of Likelihood Ratios

This section discusses the relations between the convergence of various likelihood ratios in the case where Θ is finite. It also gives some results in the infinite case.

Let us return to the case of a finite Θ with k elements. By definition, the Blackwell measure $m_\mathcal{E}$ of an experiment $\mathcal{E} = \{P_\theta; \theta \in \Theta\}$ is the image by the vector v of likelihood ratios of the sum $S = \sum_{\theta\in\Theta} P_\theta$. Therefore $m_\mathcal{E}$ is the sum of the distributions of $v : m_\mathcal{E} = \sum_{\theta\in\Theta} \mathcal{L}(v|P_\theta)$ taken under each P_θ.

Here $v(x) = \{f_\theta(x); \theta \in \Theta\}$ with $f_\theta = dP_\theta/dS$. Since in the transformation to the simplex the density f_θ becomes the coordinate function u_θ, the distribution of v under P_θ becomes the measure $u_\theta dm_\mathcal{E}$ that has density u_θ with respect to $m_\mathcal{E}$.

In many statistical problems, one needs other likelihood ratios. For instance, taking a special point $s \in \Theta$, one would want the likelihood ratios $\{dP_\theta/dP_s; \theta \in \Theta\}$ and their distribution under P_s.

Now the Radon-Nikodym density dP_θ/dP_s is well defined except on sets of P_s measure zero. We shall take for dP_θ/dP_s only the density with respect to P_s of the part of P_θ that is dominated by P_s.

On the simplex $U(\Theta)$ this becomes equal to u_θ/u_s defined where $u_s > 0$. This follows from the relations $dP_\theta = f_\theta dS$ and $dP_s = f_s dS$.

For distributions of finite-dimensional vectors (in $\mathcal{R}^k$) there are a number of definitions of convergence, all equivalent (topologically) to what is called "vague" convergence, or "ordinary" convergence, i.e., the convergence of integrals of continuous functions with compact support.

The basic definition is that the $\{P_\nu\}$ converges to a limit P_0 if $\int f dP_\nu \to \int f dP_0$ for every *bounded continuous* function f.

It can be shown (essentially as in Lemma 4 of Section 2.2) that this mode of convergence is equivalent to the convergence of the dual Lipschitz distance $\|P_\nu - P_0\|_D$ to zero.

It is also equivalent to convergence according to the Prohorov metric $\pi(P_\nu, P_0) \to 0$. The Prohorov distance $\pi(P_\nu, P_0)$ is the infimum of the number $\epsilon \in [0,1]$ such that $P_\nu(B) \leq P_0(B^\epsilon) + \epsilon$ where B^ϵ is the set $B^\epsilon = \{x : \text{dist}(x, B) \leq \epsilon\}$.

There are many other equivalent definitions, one of the oldest (and least flexible) being in terms of convergence of cumulative distribution functions.

Of course this old definition was meant for use when the underlying space is euclidean. The definition using Lipschitz functions and Prohorov's distance is meant for metric spaces and the "basic" one is meant for completely regular topological spaces. For their equivalence, when they apply, see Billingsley [1968] or Dudley [1989].

This gives the following lemma, in which one can replace the term *sequence* by net, or direction, or filter, if that seems necessary for topological purposes.

Lemma 5. *Let $\mathcal{E}_\nu = \{P_{\theta,\nu}; \theta \in \Theta\}$ be a sequence of experiments* *all indexed by a finite set Θ. Then the following are equivalent (for a limit experiment $\mathcal{F} = Q_\theta; \theta \in \Theta$):*

(i) The Blackwell measures $m_\mathcal{E}$ on $U(\Theta)$ converge in the ordinary sense (or in the sense of the dual Lipschitz distance) to that of $\mathcal{F}$.

(ii) The distributions of the likelihood ratio vector $\{dP_{\theta,\nu}/dS_\nu; \theta \in \Theta\}$ for each $s \in \Theta$ converges to the corresponding distribution of $\mathcal{F}$.

(iii) For each $s \in \Theta$, the distribution under $P_{s,\nu}$ of the likelihood ratio vector $\{dP_{\theta,\nu}/dP_{s,\nu}; \theta \in \Theta\}$ converges to the corresponding distribution for $\mathcal{F}$.

This follows almost immediately from the definitions. The only point that might not be immediate involves the transformation $\{u_\theta; \theta \in \Theta\} \rightsquigarrow \{u_\theta/u_s; \theta \in \Theta\}$ to obtain statement (iii). This transformation is not continuous on $U(\Theta)$, but for each $s \in \Theta$ its set of discontinuities has measure zero for the image of Q_s. Using the *basic* definition of convergence of positive measures one sees easily that $\mu_\nu \to \mu$, that is, $\int f d\mu_\nu \to \int f d\mu$ for bounded continuous functions, implies

$$\liminf \int f d\mu_\nu \geq \int f d\mu$$

for lower semicontinuous functions and a corresponding statement for upper semicontinuous functions.

A function whose set of discontinuities has μ measure zero coincides, except on that set, with a lower semicontinuous function, and an upper semicontinuous one. For instance if g has a set of continuity C, one can take the pointwise supremum γ of the continuous functions w such that $w(x) \leq g(x)$ on C. Similarly, let ℓ be the pointwise infimum of the continuous w such that $g(x) \geq w(x)$ on C. One has $\ell \leq g \leq \gamma$ everywhere with equality on C. If the complement of C has measure zero for the limit measure, μ, this implies $\lim \int g d\mu_\nu = \int g d\mu$. With this remark, the equivalence of statements (*ii*) and (*iii*) is readily established.

All of this applies to parameter sets Θ that are finite. Interesting experiments often correspond to infinite parameter sets. For a simple example, consider the experiment where Θ is the line and P_θ is the gaussian measure of variance unity and expectation θ. More complicated examples occur in nonparametric problems.

For the case where Θ is infinite we have only weaker results. One simple but useful remark is as follows. Let $\mathcal{E} = \{P_\theta; \theta \in \Theta\}$ and $\mathcal{F} = \{Q_\theta; \theta \in \Theta\}$ be two experiments indexed by the same finite or infinite set Θ. Suppose that $\mathcal{E}$ is dominated by some convergent sum $\mu = \sum_t c_t P_t$, $c_t \geq 0$, $\sum_t c_t = 1$ and that $\mathcal{F}$ is dominated by the corresponding sum $\nu = \sum_t c_t Q_t$.

One can then form the corresponding likelihood processes $X = \{X(t); t \in \Theta\}$ and $Y = \{Y(t); t \in \Theta\}$ where $X(t) = dP_t/d\mu$ and $Y(t) = dQ_t/d\nu$, each being given the distribution induced by the denominator measure μ.

Lemma 6. *Suppose that there is some coupling on a common probability space giving versions of the processes* $\{X(t); t \in \Theta\}$ *and* $\{Y(t); t \in \Theta\}$ *so that*

$$\sup_t \frac{1}{2} E|X(t) - Y(t)| \leq \epsilon.$$

Then $\Delta(\mathcal{E}, \mathcal{F}) \leq \epsilon$.

Proof. This is clear from (the easy part of) Theorem 2 in Section 2.2. □

Note well that the order $\sup_t E$ with the expectation is taken first. This is reminiscent of what happens for the Prohorov distance between measures. Take the space $\mathcal{X}$ of bounded numerical functions on Θ with its uniform norm $\|x\| = \sup_t |x(t)|$. Let μ and ν be two probability measures on the Borel sets of $\mathcal{X}$. According to a theorem of Strassen [1965], the Prohorov distance $\pi(\mu, \nu)$ between μ and ν does not exceed ϵ if and only if one can find a coupling on a common probability space and random processes X and Y such that $\mathcal{L}(X) = \mu$, $\mathcal{L}(Y) = \nu$ and

$$E\{1 \wedge \sup_t |X(t) - Y(t)|\} \leq 2\epsilon.$$

There is a similar theorem for dual Lipschitz distances.

Here one first takes the supremum, then the expectation. This can be a vastly stronger requirement than the requirement where the expectation is taken first.

Still, we have been unable to prove, even when Θ is finite, that couplings of the kind used in Lemma 6 always exist. (We have no counterexamples.)

When Θ is infinite, another form of convergence of experiments turns out to be useful. Let $\mathcal{E}_n$ be a sequence of experiments indexed by a parameter set Θ. Let $\mathcal{F}$ be another such experiment.

One says that $\mathcal{E}_n$ converges weakly to $\mathcal{F}$ if for each *finite* subexperiment $F \subset \Theta$, the experiments $\mathcal{E}_n(F)$ where θ is restricted to F are such that $\Delta[\mathcal{E}_n(F), \mathcal{F}(F)] \to 0$.

Similarly, Le Cam and Yang [1988] used a definition where two sequences $\mathcal{E}_n = \{P_{\theta,n}; \theta \in \Theta\}$ and $\mathcal{F}_n = \{Q_{\theta,n}; \theta \in \Theta\}$ are weakly asymptotically equivalent if for set $F_n \subset \Theta_n$ of fixed finite cardinality the distances $\Delta[\mathcal{E}_n(F_n), \mathcal{F}_n(F_n)]$ tend to zero.

These are very weak forms of convergence but they are enough to imply various results. As will be seen in Chapter 6, they are all that is needed for the Hájek–Le Cam asymptotic minimax theorem or convolution theorem.

2.5 Historical Remarks

The idea of developing statistical procedures that minimize an expected loss goes back to Laplace [1810, a, b]. He proposed to minimize expected absolute deviation. Gauss [1821], in his memoir on least squares, preferred to use square error loss $(\hat{\theta} - \theta)^2$ as more convenient and no more arbitrary than Laplace's $|\hat{\theta} - \theta|$. The idea reappears in papers of Edgeworth [1908, 1909]. According to Neyman [1952, p. 228] in his Lectures and Conferences: "After Edgeworth, the idea of the loss function was lost from sight for more than two decades...." It was truly revived only by the appearance on the statistical scene of Wald [1938]. Wald's books *Sequential Analysis* [1947] and *Statistical Decision*

Functions [1950] are based on that very idea and the idea of describing experiments by families of probability measures either on one given σ-field or on sequences of σ-fields to be chosen by the statistician. The idea seems logical enough if one is used to it. However, there is a paper by Fisher [1956] where he seems to express the opinion that such concepts are misleading and good enough only for Russian or American engineers.

The idea of comparing experiments through their sets of risk functions occurs in a famous unpublished Rand memorandum of Bohnenblust, Shapley and Sherman [1949]; they give credit for it to von Neumann. The problem was further considered by Stein [1951] and Blackwell [1951, 1953], who proved that an experiment $\mathcal{E}$ is better than an experiment $\mathcal{F}$ if and only if the canonical measure $m_{\mathcal{E}}$ of $\mathcal{E}$ is obtainable from that of $\mathcal{F}$ by a dilation (the Blackwell, Sherman, Stein theorem). For many results connected with this approach, see Strassen [1965]. Blackwell's proof, reproduced in Blackwell and Girshick [1954] for the finite case, also shows that if $\mathcal{E}$ is better than $\mathcal{F}$, one can construct a pairing (X, Y) where X would be observed in the experiment $\mathcal{E}$ and Y in the experiment $\mathcal{F}$ but where in the joint observation (X, Y) the variable X itself is a sufficient statistic.

The introduction of what is called here deficiency and of distance between experiments occurs in Le Cam [1964]. A version of this paper was actually written in late 1958 and early 1959. It took a long time to get it published, but even that did not prevent errors; Theorem 5 of that paper is wrong, as shown by Fremlin [1978]. We presume that such ideas of distances had occurred to other authors, but we do not have exact references except for the possibilities described in Stein [1951]. The equivalence of two definitions of distance–one through differences in available risk functions, the other in terms of randomizations–is in Le Cam [1964]. The exact characterization of Bayes risks in the finite-parameter case is due to Torgersen [1970]. The equivalence, in the finite-parameter case, of convergence in the sense of our distance and of convergence of distributions of likelihood ratios was given by Le Cam [1969] and Torgersen [1970]. For infinite-parameter sets the situation is more complex. See Lindae [1972] and the remarks in the this chapter.

3

Contiguity – Hellinger Transforms

We saw, in Chapter 2, that when Θ is finite, convergence of experiments in the sense of our distance is equivalent to convergence in distribution of likelihood ratios. Here we shall describe, in Section 3.1, some consequences of a condition, called *contiguity*, that simplifies many arguments in passages to the limit. Contiguity is simply an asymptotic form of absolute continuity. Theorem 1 establishes several equivalent forms of the conditions for contiguity. One of the most useful consequences of contiguity is its application to the joint limiting distribution of statistics and likelihood ratios, described in Proposition 1. We also introduce, in Section 3.2, a technical tool, the Hellinger transform, that is often convenient in studies involving independent observations. In that case it provides the same flexibility as that given by characteristic functions in the study of sums of independent random variables.

3.1 Contiguity

Consider a sequence $\{\mathcal{E}_n; n = 1, 2 \cdots\}$ of binary experiments $\mathcal{E}_n = (P_{0,n}, P_{1,n})$ where the $P_{i,n}$ are probability measures on spaces $\mathcal{X}_n$ carrying σ-fields $\mathcal{A}_n$. Binary means that the parameter space Θ consists of two points, called 0 and 1 here.

We shall first use a transformation to the interval $[0, 1]$, a slight modification of that used in Section 2.2 to obtain the Blackwell representation. It makes the spaces $(\mathcal{X}_n, \mathcal{A}_n)$ disappear, but we shall still refer to them from time to time for the convenience of the reader and other purposes. Note, however, that the properties to be discussed are properties of *distributions* of likelihood ratios and the transformations are meant to emphasize that point.

Let $S_n = P_{0,n} + P_{1,n}$ and let

$$\frac{dP_{i,n}}{dS_n} = f_{i,n}, \quad \text{for } i = 0, 1.$$

Note that the densities $f_{i,n}$ are well defined almost everywhere S_n, since S_n dominates both $P_{i,n}$. The image of S_n by the map $f_{1,n}$ is a certain measure μ_n on $[0,1]$. This is the sum of the distributions of the ratio $f_{1,n}$ for the $P_{i,n}$. The image of $P_{0,n}$ under $f_{1,n}$ has a density $1-u$ with respect to μ_n for $u \in [0,1]$. The image of $P_{1,n}$ has density u and one has $\int u d\mu_n = \int f_{1,n} dS_n = 1$, similarly for $1-u$.

Statisticians often work with *logarithms* of likelihood ratios, especially when studying independent observations. To do that, map $[0,1]$ to $[-\infty, +\infty]$ by $u \rightsquigarrow \log(u/(1-u))$. This maps μ_n to a measure M_n on $[-\infty, +\infty]$. It is the sum of the distributions of $\log(f_{1,n}/f_{0,n})$ under the $P_{i,n}$. One then has new densities

$$u \rightsquigarrow p_1(z) = \frac{e^z}{1+e^z},$$

$$1 - u \rightsquigarrow p_0(z) = \frac{1}{1+e^z},$$

with the usual conventions for infinite z.

The "usual conventions" means that the loglikelihood $\log(u/(1-u))$ may actually take the values $-\infty$ and $+\infty$. This is meant to be and is used in the proof of Theorem 1.

If pulled back to the original space $(\mathcal{X}_n, \mathcal{A}_n)$, the ratio $u/(1-u)$ would become $f_{1,n}/f_{0,n}$. This can be taken as a definition for the likelihood ratio of $P_{1,n}$ to $P_{0,n}$. Its logarithm can take both values $-\infty$ and $+\infty$. That feature creates some inconvenience when adding the logarithms of likelihood ratios for independent observations. Thus, we shall most often use a different definition. Let $P'_{1,n}$ be the part of $P_{1,n}$ that is dominated by $P_{0,n}$. Define a function Λ_n on the support of $P_{0,n}$ by

$$\Lambda_n = \log \frac{dP'_{1,n}}{dP_{0,n}}. \tag{3.1}$$

In the sequel, we write, for simplicity, $\Lambda_n = \log(dP_{1,n}/dP_{0,n})$. It will always be meant in the sense of (3.1).

One can define statistics T_n, measurable maps from $(\mathcal{X}_n, \mathcal{A}_n)$ to the line. We shall call them "statistics available on the experiments $\mathcal{E}_n$." One says that $T_n \to 0$ in $P_{0,n}$-probability if $\int \min(1, |T_n|) dP_{0,n} \to 0$ as $n \to \infty$.

One says that the sequence of distributions $\{\mathcal{L}(T_n|P_{0,n}); n = 1, 2, \cdots\}$ is relatively compact (on the line) if for every $\epsilon > 0$ there is a $b(\epsilon)$ and an integer $N(\epsilon)$ such that $n \geq N(\epsilon)$ implies $P_{0,n}\{|T_n| > b(\epsilon)\} < \epsilon$. Actually this is a definition of tightness of distributions of the T_n under the $P_{0,n}$, but here, as in all complete separable metric spaces, the tightness of sequences of Borel measures is equivalent to relative compactness. From such sequences one can extract subsequences converging to limits that are probability measures on the line. Such limits of subsequences are called cluster points. The set of all cluster points of the sequence $\{\mathcal{L}(T_n|P_{0,n})\}$ is the set $C = \bigcap_n C_n$ with C_n equal to the closure of $\{\mathcal{L}(T_m|P_{0,m}); m \geq n\}$.

With these notations one can assert the following, with all limits taken as $n \to \infty$.

Theorem 1. *Let $\{M_n; n = 1, 2, \cdots\}$ be the sequence of images of $S_n = P_{0,n} + P_{1,n}$ on $[-\infty, +\infty]$. Let $F_n(dz) = p_0(z)M_n(dz)$ be the distribution under $P_{0,n}$ of the log likelihood ratio and let T_n denote real-valued statistics available on the experiment $\mathcal{E}_n = (P_{0,n}, P_{1,n})$. The following seven properties are all equivalent:*

(a) *Every sequence $\{T_n\}$ that tends to zero in $P_{0,n}$ probability also tends to zero in $P_{1,n}$ probability.*

(b) *For every sequence $\{T_n\}$ such that $\{\mathcal{L}(T_n|P_{0,n})\}$ is relatively compact, $\{\mathcal{L}(T_n|P_{1,n})\}$ is also relatively compact.*

(c) *If M is a cluster point of the sequence $\{M_n\}$ on $[-\infty, +\infty]$ then $M[\{+\infty\}] = 0$.*

(d) *Let $dF_n = p_0 dM_n$. If F is a cluster point of the sequence $\{F_n\}$ on $[-\infty, +\infty]$, then $\int e^z F(dz) = 1$.*

(e) *If* $z_n \in [-\infty, +\infty)$ *is such that* $z_n \to +\infty$ *then* $M_n\{[z_n, \infty]\} \to 0$.

(f) *For every* $\epsilon > 0$ *there exist real numbers* $N(\epsilon)$ *and* $b(\epsilon)$ *such that* $M_n\{[b(\epsilon), \infty]\} < \epsilon$ *for all* $n \geq N(\epsilon)$.

(g) *If* $\mathcal{E} = (P_0, P_1)$ *is a cluster point of the sequence of experiments* $\mathcal{E}_n$*, then* P_1 *is dominated by* P_0.

Proof. (1) $[c \Leftrightarrow d]$. If (c) holds, there is a subsequence $\{M_{n(j)};\ j = 1, 2, \cdots,\}$ of the sequence $\{M_n\}$ such that $M_{n(j)} \to M$ weakly. This means that for every continuous real-valued function ϕ defined on $[-\infty, +\infty]$ the integrals $\int \phi dM_{n(j)}$ converge to $\int \phi dM$.

Since p_0 defined by $p_0(z) = 1/(1 + e^z)$ is continuous on $[-\infty, +\infty]$ the integrals $\int \phi p_0 dM_{n(j)}$ converge to $\int \phi p_0 dM$. Therefore $F_{n(j)}$ converges to a limit F such that $dF = p_0 dM$. However p_0 vanishes at $+\infty$. Therefore F has no mass at $+\infty$ and

$$\int e^z F(dz) = \int_{z<+\infty} e^z F(dz) = \int_{z<+\infty} p_1 dM = 1 - M[\{+\infty\}].$$

Thus $\int e^z F(dz) = 1$ if and only if $M[\{+\infty\}] = 0$.

(2) $[f \Leftrightarrow c]$. Let M be a cluster point of the sequence $\{M_n\}$, limit of a subsequence $\{M_{n(j)}, j = 1, 2, \cdots,\}$. Let ϕ_b be the continuous function equal to zero for $z < b$, to one for $z > b+1$, and defined by linear interpolation in $[b, b+1]$. Then $\int \phi_b dM_{n(j)} \to \int \phi_b dM$. Thus, if (f) holds there is a b such that eventually $\int \phi_b dM_{n(j)} < \epsilon$ and therefore $\int \phi_b dM \leq \epsilon$. Thus $\lim_{b\to\infty} \int \phi_b dM = M[\{+\infty\}] \leq \epsilon$. Since ϵ is arbitrary, one has $M[\{+\infty\}] = 0$.

Conversely if (f) does not hold, there is an $\epsilon > 0$ such that for each j there is some $n(j) \geq j$ such that $M_{n(j)}\{[j, \infty]\} \geq \epsilon$. Thus $\lim_j \int \phi_b dM_{n(j)} = \int \phi_b dM \geq \epsilon$ for each $b < \infty$. This implies $\lim_b \int \phi_b dM = M[\{+\infty\}] \geq \epsilon$, contradicting (c).

(3) $[e \Leftrightarrow f]$. This is quite clear.

(4) $[f \Leftrightarrow a]$. Assume that the sequence $\{T_n\}$ tends to zero in $P_{0,n}$ probability. This means that $\int \min[1, |T_n|] dP_{0,n} \to 0$. Let

$\Lambda_n = \log(dP_{1,n}/dP_{0,n})$, where the Radon-Nikodym density is that of the part of $P_{1,n}$ that is dominated by $P_{0,n}$.

Here it is convenient to partition the space $\mathcal{X}_n$ as $\mathcal{X}_n = A_n \cup B_n$ where A_n carries $P_{0,n}$ and B_n carries the part of $P_{1,n}$ that is $P_{0,n}$ singular. Let Λ_n^* be defined in the same way as Λ_n but equal to ${\Lambda_n}^* = +\infty$ on the subset B_n. Then

$$M_n[b, \infty] = P_{0,n}[\Lambda_n \geq b] + P_{1,n}[{\Lambda^*}_n \geq b].$$

Also

$$\int \min[1, |T_n|] dP_{1,n} \leq \int \min[1, |T_n|] I[\Lambda_n < b] e^{\Lambda_n} dP_{0,n} + P_{1,n}[\Lambda_n^* \geq b].$$

Since $I[\Lambda_n < b]e^{\Lambda_n}$ is bounded, the integral on the right tends to zero. The second term on the right is bounded by $M_n\{[b, \infty]\}$. Property (f) says that this can be made as small as one pleases. The result follows.

(5) $[a \Leftrightarrow e]$. Let $\{z_n\}$ be a sequence such that $z_n \to \infty$. Let $T_n = I[z_n, \infty]$. Then

$$\int T_n dP_{0,n} = \int_{z_n}^{\infty} \frac{1}{(1+e^z)} dM_n \leq \frac{2}{1+e^{z_n}} \to 0 \quad \text{as } n \to \infty.$$

According to (a)

$$\int T_n dP_{1,n} = \int_{z_n}^{\infty} \frac{e^z}{1+e^z} dM_n \to 0.$$

Thus $M_n\{[z_n, \infty]\} \to 0$.

(6) $[f \Leftrightarrow b]$. For real variables T_n to say that $\{\mathcal{L}[|T_n|P_{0,n}]\}$ is relatively compact is to say that for each $\epsilon > 0$ there is an $a(\epsilon)$ such that $P_{0,n}[|T_n| \geq a(\epsilon)] \leq \epsilon$ for all n. For any $b > 0$ one has

$$P_{1,n}[|T_n| \geq a] \leq e^b P_{0,n}[|T_n| \geq a] + M_n[b, \infty].$$

Select b according to (f) so that $M_n\{[b, \infty]\} < \epsilon/2$. Then select a so that $P_{0,n}[|T_n| \geq a] < (\epsilon/2)e^{-b}$. This shows that $(f) \Rightarrow$

(b). To get the reverse implication $[b \Rightarrow f]$, let $T_n = \Lambda_n = dP_{1,n}/dP_{0,n}$ on the subset that carries $P_{0,n}$ and put $T_n = n$ otherwise. Such a sequence T_n is relatively compact for $P_{0,n}$ by Markov's inequality, hence also for $P_{1,n}$ if (b) holds. This implies (f).

This concludes the proof of the theorem except for the part that involves (g). For that part, let us first make some observations. The experiment $\mathcal{E}_n = (P_{0,n}, P_{1,n})$ is equivalent to its Blackwell representation. This is clearly equivalent to the experiment $(F_{0,n}, F_{1,n})$ where $dF_{i,n} = p_i dM_n$ on $[-\infty, +\infty]$. The convergence of $\mathcal{E}_n$ is equivalent to the convergence on $[-\infty, +\infty]$ of $F_{i,n}$ to limits F_i, $i = 1, 2$. Now let us show the following.

(7) $[g \Leftrightarrow c]$. The condition $M[\{+\infty\}] = 0$ is equivalent to the condition that F_1 be dominated by F_0 since the density of F_0 vanishes only at the point $+\infty$. Note also that if (F_0, F_1) and (P_0, P_1) are equivalent experiments, P_1 is dominated by P_0 if and only if F_1 is dominated by F_0 by Theorem 1 of Section 2.3. Indeed, if P_1 had a singular part with mass $\alpha > 0$ with respect to P_0 then there would be a test ϕ of P_1 against P_0 with risks $\int \phi dP_0 = 0$ and $\int (1-\phi) dP_1 \leq 1-\alpha$ and similarly, interchanging (F_0, F_1) and (P_0, P_1). □

Definition 1. *If a sequence $\mathcal{E}_n = (P_{0,n}, P_{1,n})$ of binary experiments satisfies one of (and therefore all) the conditions(a)–(g) of Theorem 1, one says that the sequence $\{P_{1,n}\}$ is contiguous to $\{P_{0,n}\}$. If the conditions are also satisfied when $P_{0,n}$ and $P_{1,n}$ are interchanged, one says that the sequences $\{P_{0,n}\}$ and $\{P_{1,n}\}$ are contiguous.*

The symmetric version of this definition was introduced by Le Cam [1960]. The one-sided version was used by Hájek [1962] and Hájek and Šidák [1967]. Note, however, that in these works and many others one typically assumes that under $P_{0,n}$ the loglikelihood ratios $\Lambda_n = \log(dP_{1,n}/dP_{0,n})$ have on $[-\infty, +\infty]$ a limiting distribution carried by $(-\infty, +\infty)$. This means that the limit M of condition (c) is assumed to be such that $M[\{-\infty\}] = 0$. This is the condition symmetric to the $M[\{+\infty\}] = 0$ of (c). Thus if

$\mathcal{L}[\Lambda_n|P_{0,n}]$ has a limit on $(-\infty, +\infty)$ and if $\{P_{1,n}\}$ is contiguous to $\{P_{0,n}\}$, it follows that $\{P_{0,n}\}$ is also contiguous to $\{P_{1,n}\}$.

This symmetric property is, of course, due to the fact that $\mathcal{L}[\Lambda_n|P_{0,n}]$ has a limit on $(-\infty, +\infty)$. It does not hold otherwise. An example can be constructed by taking the joint distributions of independent variables $(X_1, X_2, ..., X_n)$ where under $P_{0,n}$ the individual distributions are uniform on $[0, 1]$ and under $P_{1,n}$ they are uniform on $[1/n, 1]$.

We have stated Theorem 1 for *sequences*; it does not hold as such for *nets* or *filters*. Examples and the necessary modifications can be found in Le Cam [1986, p. 87].

Remark. The conditions of Theorem 1 have been stated in terms of *real-valued* random variables T_n available on $\mathcal{E}_n$. They could also be stated for variables T_n that take values in any fixed complete separable metric space, that is, for stochastic processes. Condition (a) would then be replaced by "if T_n tends for $P_{0,n}$ to a constant a, it also tends in $P_{1,n}$ probability to the same constant." For condition (b) and relative compactness see the more general results of Le Cam [1986, p. 90].

One of the most useful consequences of the contiguity conditions is the following, in which $\Lambda_n = \log(dP_{1,n}/dP_{0,n})$, is defined by (3.1).

Proposition 1. *Let $\{T_n\}$ be a sequence of random variables with T_n available on $\mathcal{E}_n = (P_{0,n}, P_{1,n})$. Assume that*

(i) $\{P_{1,n}\}$ is contiguous to $\{P_{0,n}\}$ and

(ii) the joint distributions $\mathcal{L}\{(T_n, \Lambda_n)|P_{0,n}\}$ tend to a limit F on
$(-\infty, +\infty) \times [-\infty, +\infty]$.

Then $\mathcal{L}\{(T_n, \Lambda_n)|P_{1,n}\}$ tend to a limit G and $G(dt, d\lambda) = e^{\lambda}F(dt, d\lambda)$.

Note. The result is also valid for variables T_n that are vector valued or variables that take values in Polish spaces. For more general results see Le Cam [1986, p. 90].

Proof. Let ϕ be a bounded continuous function of (t, λ) on $(-\infty, +\infty) \times [-\infty, +\infty]$.

Let $F_{i,n} = \mathcal{L}[(T_n, \Lambda_n)|P_{i,n}]$. One can write

$$\int \phi dF_{1,n} = \int \phi(t, \lambda) e^{\lambda} F_{0,n}(dt, d\lambda) + \epsilon_n$$

where ϵ_n is the contribution of the part of $P_{1,n}$ that is singular with respect to $P_{0,n}$. According to the contiguity property, $\epsilon_n \to 0$. The remaining integral can be written

$$\int \phi(t, \lambda) e^{\lambda} F_{0,n}(dt, d\lambda) = \int \phi(t, \lambda) \min[b, e^{\lambda}] F_{0,n}(dt, d\lambda)$$
$$+ \int \phi(t, \lambda)[e^{\lambda} - b]^{+} F_{0,n}(dt, d\lambda).$$

For every $\epsilon > 0$ one can select a b such that the second integral on the right is eventually less than ϵ in absolute value. The first term on the right tends to $\int \phi(t, \lambda) \min[b, e^{\lambda}] F_0(dt, d\lambda)$. The desired result follows by letting b tend to infinity. □

The foregoing applies in particular to Λ_n itself. Suppose, for example, that $\mathcal{L}[\Lambda_n|P_{0,n}]$ converges to a gaussian $F = \mathcal{N}(m, \sigma^2)$. Then under $P_{1,n}$ the distributions $\mathcal{L}(\Lambda_n|P_{1,n})$ will converge to G such that $G(d\lambda) = e^{\lambda} F(d\lambda)$. Since G must be a probability measure, or from (d) of Theorem 1, it follows that $m + (\sigma^2/2) = 0$ and $G = \mathcal{N}(m + \sigma^2, \sigma^2)$.

The equality $m = -\sigma^2/2$ that occurs here is a classical one. One finds it, for instance, in the standard treatment of maximum likelihood estimation under Cramér's conditions. There it is derived from conditions of differentiability under the integral sign.

Let's look at another particular case. Assume that $\mathcal{L}[(T_n, \Lambda_n)| P_{0,n}]$ converges to a bivariate normal distribution F with a mean vector and a covariance matrix given respectively by

$$\begin{pmatrix} a \\ b \end{pmatrix} \quad \text{and} \quad \Gamma = \begin{pmatrix} A & C \\ C & B \end{pmatrix}.$$

The logarithm ψ of the characteristic function of F has the form

$$\psi(u, v) = \log E \exp[iuT + iv\Lambda]$$

$$= iua + ivb - \frac{1}{2}[Au^2 + 2Cuv + Bv^2].$$

Thus $Ee^{\Lambda}\exp[iuT + iv\Lambda]$ is obtainable by replacing iv by $iv+1$. Carrying out the algebra, one sees that

(1) the quadratic terms in u^2, uv and v^2 are unchanged;
(2) the coefficient a is replaced by $a + C$; and
(3) the coefficient $b = -B/2$ is replaced by $b + B = B/2$.

In other words the distribution $\mathcal{L}[T_n, \Lambda_n | P_{1,n}]$ also tends to a limit. It is a bivariate normal distribution that has the same covariance structure as the limit of $\mathcal{L}[T_n, \Lambda_n | P_{0,n}]$. The vector of means is changed by adding the appropriate covariance terms of the limiting distribution, i.e., $\begin{pmatrix} a + C \\ b + B \end{pmatrix}$.

There are other situations (see Le Cam [1960]) where convergence of $\mathcal{L}[T_n, \Lambda_n | P_{0,n}]$ to F yields a limit $G = \lim \mathcal{L}[T_n, \Lambda_n | P_{1,n}]$ that is easily identifiable. For instance, if F is infinitely divisible and the corresponding marginal $\mathcal{L}(T|F)$ is gaussian, then the first marginal of G is also normal with the same variance as in F and a mean $a + C$ where C is the covariance term in the normal component of F.

The preceding Proposition 1 can be applied to simplify the criteria for convergence of experiments given in Chapter 2. There we considered likelihood ratio vectors $V_{n,\theta} = \{dP_{t,n}/dP_{\theta,n};\ t \in \Theta\}$ and required that for *each* $\theta \in \Theta$ (finite) the distributions $\mathcal{L}[V_{n,\theta} | P_{\theta,n}]$ converge.

If it happens that all $\{P_{\theta,n}\}$ are contiguous to a $\{P_{s,n}\}$ for a particular $s \in \Theta$, then it is sufficient to check that $\mathcal{L}\{V_{n,s} | P_{s,n}\}$ converge for *that particular* s. This is an easy consequence of Proposition 1 applied to the pairs (T_n, Λ_n), where T_n is the vector $V_{n,\theta}$ and Λ_n is $\log(dP_{\theta,n}/dP_{s,n})$.

3.2 Hellinger Distances, Hellinger Transforms

One of the fundamental distances between finite signed measures is given by the L_1-norm $\|\mu\|_1 = \sup_f |\int f d\mu|$ where f runs through all measurable functions such that $|f| \leq 1$. It was already used in Chapter 2 for the definition of deficiencies. We shall often use it without the subscript "1," simply as $\|\mu\|$. It is usually much larger than the dual Lipschitz norm $\|\mu\|_D$, where f is further restricted by the condition that $|f(x) - f(y)| \leq \text{dist}(x, y)$. Neither of these distances behaves comfortably when passing to direct products, as encountered in the study of independent observations. Another distance, the Hellinger distance, behaves in a much better manner in such cases. If P and Q are positive measures on a σ-field $\mathcal{A}$ the Hellinger distance between the two is $h(P, Q)$ defined by

$$h^2(P,Q) = \frac{1}{2}\int(\sqrt{dP} - \sqrt{dQ})^2.$$

For probability measures this can also be written

$$h^2(P,Q) = 1 - \rho(P,Q),$$

where $\rho(P, Q)$ is the affinity $\rho(P,Q) = \int \sqrt{dPdQ}$. The reader who is not familiar with a symbol such as $\int \sqrt{dPdQ}$ can interpret it as $\int \sqrt{dPdQ} = \int \sqrt{fg}d\mu$ where $f = dP/d\mu$, $g = dQ/d\mu$ for any measure μ that dominates both P and Q. The value of h^2 does not change with the dominating measure μ. That justifies the use of notation $\int \sqrt{dPdQ}$. The Hellinger distance was not actually used by Hellinger, as far as we know. Hellinger did not have the benefit of the Radon-Nikodym theorem and proceeded to a direct definition. If one mimics it for $\int \sqrt{dPdQ}$, it would look as follows. Take a finite partition
$\{A_1, A_2, \ldots, A_k\}$ of the underlying space. Compute

$$\sum[P(A_j)Q(A_j)]^{\frac{1}{2}}, j = 1, 2, \ldots, k.$$

Pass to the limit as the partition is refined indefinitely. In the case of the affinity, convexity properties ensure that the limit exists and could be replaced by an infimum taken over all measurable finite partitions. The use of the Hellinger affinity in statistics may be credited to Kakutani [1948]; see also Kraft [1955]. A related distance was used for a long time in quantum mechanics, because probabilities there are given by the absolute square of wave functions.

It is easy to check that for probability measures P and Q, the L_1-norm and the Hellinger distance are related by the following inequalities:

$$\begin{aligned} h^2(P,Q) &\leq \frac{1}{2}\|P-Q\|_1 \\ &\leq h(P,Q)\sqrt{2-h^2(P,Q)} = [1-\rho^2(P,Q)]^{1/2}. \end{aligned}$$

To prove this, let $\mu = P+Q$, $f = dP/d\mu$, $g = dQ/d\mu$. Then

$$\begin{aligned} h^2(P,Q) &= \frac{1}{2}\int(\sqrt{f}-\sqrt{g})^2 d\mu \\ &\leq \frac{1}{2}\int|\sqrt{f}-\sqrt{g}|\ |\sqrt{f}+\sqrt{g}|d\mu \\ &= \frac{1}{2}\int|f-g|\ d\mu. \end{aligned}$$

Similarly, by Schwarz's inequality

$$\frac{1}{2}\int|\sqrt{f}-\sqrt{g}|\ |\sqrt{f}+\sqrt{g}|d\mu$$

$$\leq \{\frac{1}{2}\int(\sqrt{f}-\sqrt{g})^2 d\mu\}^{1/2}\ \{\frac{1}{2}\int(\sqrt{f}+\sqrt{g})^2 d\mu\}^{1/2}$$

and

$$\frac{1}{2}\int(\sqrt{f}+\sqrt{g})^2 d\mu = 1+\int\sqrt{fg}d\mu = (1+\rho) = (2-h^2).$$

It is also easy to check that $\|P\wedge Q\|_1 = 1-(1/2)\|P-Q\|_1$ is the sum of errors for a test between P and Q that minimizes the sums of errors. Thus $(1/2)\|P-Q\|_1$ has a specific statistical meaning. Since $h(P,Q)$ is related to it by the inequalities written above one can often work with $h(P,Q)$ instead of $(1/2)\|P-Q\|_1$. Now $h(P,Q)$, or more specifically $\rho(P,Q)$, behaves very well in product situations. Suppose, for instance, that P_0 and P_1 are two probability measures on a σ-field $\mathcal{A}$ product of σ-fields $\mathcal{A}_j$, $j \in J$ (J need not be a finite set). Suppose that P_i is the product measure $\Pi_j p_{i,j}$, where the $p_{i,j}$ are possible distributions of independent observations ξ_j. Then the affinity between P_0 and P_1 is

$$\rho(P_0,P_1) = \Pi_{j\in J}\rho(p_{0,j},p_{1,j}).$$

This is easy to check since $\rho(P_0, P_1)$ can be written as the multiple integral

$$\rho(P_0, P_1) = \int \Pi_j \{dp_{0,j}\, dp_{1,j}\}^{1/2}.$$

In other terms if $h_j = h(p_{0,j}, p_{1,j})$, one will have

$$1 - h^2(P_0, P_1) = \Pi_j(1 - h_j^2) \leq \exp\{-\sum_j h_j^2\}.$$

These relations are often very convenient; we shall come back to their use in Chapter 5. For now, note that if we make all $p_{i,j}$ depend on an integer n, getting $P_{i,n} = \Pi_{j \in J_n} p_{i,j,n}$, a statement that $\Pi_j(1 - h_{j,n}^2)$ remains bounded away from zero is equivalent to the statement that $P_{0,n}$ and $P_{1,n}$ do not separate entirely. That is, the sum of error probabilities $\|P_{0,n} \wedge P_{1,n}\|$ stays away from zero. A statement that the affinity $\Pi_j(1 - h_{j,n}^2)$ stays away from unity is equivalent to the statement that the experiments $(P_{0,n}, P_{1,n})$ stay away from the trivial experiment $\mathcal{E} = (P, P)$ where the two possible measures P_0 and P_1 are one and the same, equal to P.

The affinity $\int \sqrt{dPdQ}$ is a special value of what is called the*Hellinger transform*. This is defined as follows.

Let $\mathcal{E} = \{P_\theta; \theta \in \Theta\}$ be an experiment indexed by the set Θ. Let $\{\alpha_\theta;\, \theta \in \Theta\}$ be numbers such that (i) $\alpha_\theta \geq 0$, (ii) $\sum_\theta \alpha_\theta = 1$, and (iii) only a finite number of the α_θ are strictly positive. The Hellinger transform of $\mathcal{E}$ evaluated at $\{\alpha_\theta;\, \theta \in \Theta\}$ is

$$\int \Pi_\theta (dP_\theta)^{\alpha_\theta}.$$

Here again this integral may be interpreted as $\int \Pi_\theta f_\theta^{\alpha_\theta} d\mu$ where μ is any positive measure that dominates all the P_θ for which $\alpha_\theta > 0$ and where $f_\theta = dP_\theta / d\mu$.

The use of Hellinger transforms is similar to the use of Fourier or Laplace transforms: *It transforms the operation of taking direct products of experiments into pointwise multiplication of the Hellinger transforms.* Indeed, let $\mathcal{E} = \{P_\theta; \theta \in \Theta\}$ and $\mathcal{F} = \{Q_\theta;\, \theta \in \Theta\}$ be two experiments indexed by the same Θ. Let $\mathcal{E} \otimes \mathcal{F}$

be the experiment that consists in performing $\mathcal{E}$ and then performing $\mathcal{F}$ independently of what happened in $\mathcal{E}$. We shall call $\mathcal{E} \otimes \mathcal{F}$ the direct product of the experiments $\mathcal{E}$ and $\mathcal{F}$. The measures in $\mathcal{E} \otimes \mathcal{F}$ are the product measures $P_\theta \otimes Q_\theta$. It follows that if $\phi_\mathcal{E}$ and $\phi_\mathcal{F}$ are the respective transforms of $\mathcal{E}$ and $\mathcal{F}$ then $\phi_{\mathcal{E}\otimes\mathcal{F}} = \phi_\mathcal{E}\phi_\mathcal{F}$. Indeed, for a given $\alpha = \{\alpha_\theta;\ \theta \in \Theta\}$, one has

$$\int \Pi_\theta\{d[P_\theta \otimes Q_\theta]\}^{\alpha_\theta} = \int \Pi_\theta(dP_\theta)^{\alpha_\theta} \int \Pi_\theta(dQ_\theta)^{\alpha_\theta},$$

by Fubini's theorem.

This multiplicative property is the most important characteristic of Hellinger transforms. However, they also possess other usable properties. To describe some, let Θ be finite and let $U(\Theta)$ be the simplex of vectors $\alpha = \{\alpha_\theta; \theta \in \Theta\}$ such that $\alpha_\theta \geq 0$, $\sum_\theta \alpha_\theta = 1$.

Proposition 2. *The Hellinger transform* $\phi_\mathcal{E}(\alpha) = \int \Pi_\theta(dP_\theta)^{\alpha_\theta}$ *of an experiment* $\mathcal{E} = \{P_\theta\colon \theta \in \Theta\}$ *characterizes that experiment up to an equivalence. Convergence of experiments* $\mathcal{E}_n = \{P_{\theta,n};\ \theta \in \Theta\}$ *to a limit* $\mathcal{E} = \{P_\theta\colon \theta \in \Theta\}$ *in the sense of the distance* Δ *defined in Chapter 2 is equivalent to convergence of their Hellinger transforms* $\phi_{\mathcal{E}_n}$ *pointwise on* $U(\Theta)$.

Proof. Let $S = \sum_\theta P_\theta$. Let $f_\theta = dP_\theta/dS$. Then $\phi_\mathcal{E}(\alpha) = \int(\Pi_\theta f_\theta^{\alpha_\theta})\, dS$. This shows that $\phi_\mathcal{E}$ depends on $\mathcal{E}$ only through the distribution of the vector $\{f_\theta;\ \theta \in \Theta\}$. Thus $\phi_\mathcal{E}$ depends only on the type of $\mathcal{E}$.

Also, by Lemma 1, Section 2.2, convergence of experiments implies convergence of their Hellinger transforms.

To prove the converse, it is convenient to assume that the experiments are given in their Blackwell canonical representation on the simplex $U(\Theta)$. Thus, the f_θ become the coordinates in the simplex, and the problem is to characterize the sum S of the measures there.

Now take an $\alpha = \{\alpha_\theta\}$ and a particular $t \in \Theta$. Assume $\alpha_t > 0$ and let Λ_θ be $\log f_\theta/f_t$. Then

$$\phi_\mathcal{E}(\alpha) = \int \Pi_\theta\left(\frac{f_\theta}{f_t}\right)^{\alpha_\theta} dP_t = \int \exp\{\sum_\theta \alpha_\theta\Lambda_\theta\}\, dP_t.$$

For fixed t the transform $\phi_{\mathcal{E}}$ appears as a Laplace transform of the distribution under P_t of the vector $\{\Lambda_\theta; \theta \in \Theta, \theta \neq t\}$. It is given on the set $\{\alpha : \alpha_\theta > 0, \sum_{\theta \neq t} \alpha_\theta < 1\}$. Here the ratios f_θ / f_t may vanish. Nevertheless the uniqueness of Laplace transforms ensures that the part of the distribution of $\{f_\theta / f_t; \theta \neq t\}$ where all the f_θ are different from zero is well determined. That means that the sum S is well determined on the interior of $U(\Theta)$.

Remove that part of S and the corresponding part of $\phi_{\mathcal{E}}$. We are left with a measure on the faces of the simplex $U(\Theta)$. Each is subject to the same argument by putting the corresponding coordinate of α_θ to zero. Proceeding down on the faces, one sees that S is well determined.

The convergence properties can be handled in the same way, appealing to the relations between convergence of Laplace transforms and convergence of positive measures. □

Finally, let us note that, if necessary, contiguity properties can be checked on Hellinger transforms. For this take experiments $\mathcal{E}_n = (P_{0,n}, P_{1,n})$ with Hellinger transforms ϕ_n defined by

$$\phi_n(\alpha) = \int (dP_{0,n})^{1-\alpha} (dP_{1,n})^\alpha.$$

Note that if $\phi(\alpha) = \int (dP_0)^{1-\alpha} (dP_1)^\alpha$ then $\lim_{\alpha \to 0, \alpha > 0} \phi(\alpha)$ is the norm of the part of P_0 that is dominated by P_1. Indeed, let D be the set where both densities $f_i = dP_i/(dP_0 + dP_1)$ do not vanish. If $\alpha(1-\alpha) > 0$, the integral $\int (dP_0)^{1-\alpha} (dP_1)^\alpha$ is equal to $\int_D (dP_0)^{1-\alpha} (dP_1)^\alpha$, whose limit as $\alpha \to 0$ is $P_0(D)$. Similarly for α tending to 1,

$$\lim_{\alpha \to 1, \alpha < 1} \phi(\alpha) = P_1(D),$$

the norm of the part of P_1 that is dominated by P_0.

This gives the following.

Proposition 3. *Let ϕ_n be as defined above. Then the experiments $\mathcal{E}_n = (P_{0,n}, P_{1,n})$ satisfy the equivalent conditions of Theorem 1 if and only if*

$$\lim_{\alpha \to 1, \alpha < 1} \liminf_n \phi_n(\alpha) = 1.$$

This is clear from Theorem 1. Note that, because of the convexity of $\log \phi_n(\alpha)$, one has for $\alpha \in [1/2, 1]$,

$$\log \phi_n(\alpha) \geq 2\alpha \log \phi_n(\frac{1}{2}) = 2\alpha \log \rho\,(P_{0,n}, P_{1,n})$$

for the affinity ρ between $P_{0,n}$ and $P_{1,n}$. However, the condition

$$\liminf{}_n \rho\,(P_{0,n}, P_{1,n}) > 0,$$

equivalent to the fact that $\mathcal{E}_n$ stays away from the perfect experiment formed by disjoint measures, is far from implying the contiguity condition, while the conditions of Theorem 1 imply, of course, that $\liminf{}_n \rho\,(P_{0,n}, P_{1,n}) > 0$. □

3.3 Historical Remarks

Sometime in 1955 or 1956, it became imperative to assign a name to a sort of uniform asymptotic mutual absolute continuity of measures. The name selected by Le Cam, with the help of James Esary, was "contiguity." The resulting theory, including most of the results of Section 1, can be found in Le Cam [1960]. Later, Hájek [1962] found the idea attractive and useful. Roussas [1972] devoted a book to it and its applications. Since then, many authors have made efforts to prove contiguity of sequences of measures under various circumstances. See, for example, Greenwood, Shiryaev [1985], and Jacod and Shiryaev [1987].

What is called Hellinger distance or Hellinger affinity here does not seem to have been considered by Hellinger. The concepts appear in Kakutani [1948], where they were used very efficiently. They occur again, with the name of Hellinger attached to them, in Kraft's thesis [1955]. Matusita was already using the Hellinger distance and affinity in 1953. See his papers of [1955] and [1967] and the references therein.

What Hellinger considered were direct definitions of integrals of the type $\int dPdQ/dM$ with P and Q both dominated by M.

Further developments were given by Riesz [1940] and Dieudonné [1941, 1944] for general "homogeneous functions of measures." What we have called the Hellinger transform $\int(dP)^{1-\alpha}(dQ)^{\alpha}$, $\alpha \in [0, 1]$, is a particular case of such homogeneous functions.

We do not know who first introduced their use in statistics. For the Hellinger distance itself, it must have occurred early in quantum mechanics because there the probabilities are given by the square modulus of wave functions. Halphen [1957] made use of something similar to Hellinger transforms, but his aim seems different from ours. The Hellinger transform occurs explicitly in Le Cam [1969] and Torgersen [1970]. Torgersen also shows that if an experiment $\mathcal{E}$ is better than experiment $\mathcal{F}$ then the Hellinger transform of $\mathcal{E}$ is smaller than that of $\mathcal{F}$ but the converse does not hold.

4

Gaussian Shift and Poisson Experiments

4.1 Introduction

This chapter is an introduction to some experiments that occur in their own right but are also very often encountered as limits of other experiments. The gaussian ones have been used and overused because of their mathematical tractability, not just because of the Central Limit Theorem. The Poisson experiments cannot be avoided when one models natural phenomena. We start with the common gaussian experiments. By "gaussian" we shall understand throughout they are the "gaussian shift" experiments, also called *homoschedastic*. There is a historical reason for that appellation: In 1809 Gauss introduced them (in the one-dimensional case) as those experiments where the maximum likelihood estimate coincides with the mean of the observations. This fact, however, will not concern us.

After giving their definition, we show that any gaussian experiment defines a Hilbert space that can be finite- or infinite-dimensional. The underlying parameter space gets embedded as a subset of the Hilbert space. Purely geometric properties of that subset determine the statistical properties of the corresponding experiment.

We stay at an elementary level and, except for a brief mention in the section on history and in Chapter 7, we do not enter into the results due to Charles Stein, L.D. Brown, or D. Donoho.

Another kind of experiments introduced here is the Poisson experiments. Those are the experiments where the observable are random discrete measures with an independence condition for disjoint sets. The expected number of points falling in a set A gives a finitely additive measure or "content" λ such that $\lambda(A)$ is the expected number of points in A. The Poisson experiments are "parametrized" by some family of contents λ. (We use the word "content" on purpose; there are too many entities called "measures" floating around. The word in question is also meant to remind the reader that no infinite operations are performed on the content or its domain. If the reader prefers to assume that λ is a measure, no harm will ensue as long as one works on the ring of sets of finite measure.) It is shown that Poisson experiments provide a natural approximation to experiments where the observable are independent, and each individual observation gives only a little information (see Proposition 2 and Theorem 1, Section 4.3.) Infinitely divisible experiments are always weak limits of Poisson experiments.

4.2 Gaussian Experiments

Every statistician has encountered experiments $\mathcal{E} = \{P_\theta; \theta \in \Theta\}$ where Θ is a subset of a k-dimensional vector space $\Re^k$ and P_θ has, with respect to the Lebesgue measure of $\Re^k$, a density of the form:

$$[\frac{\det M_\theta}{(2\pi)^k}]^{\frac{1}{2}} \exp\{-(1/2)[x-\theta]' M_\theta [x-\theta]\}.$$

Note the occurrence of a nonsingular matrix M_θ that may vary as θ does. This would be the *heteroschedastic* case studied in multivariate analysis. Here we shall consider only the shift case where $M_\theta = M$ independent of θ. Note also the notation $x'Mx$ that presupposes the choice of a basis or an identification between the vector space $\Re^k$ and its dual. This would be inconvenient here. Because M is presumably independent of θ, it will be simpler to use the fact that $x'Mx$ is the square of a norm (or seminorm) $x'Mx = \|x\|^2$.

It is also inconvenient to force the parameter θ to play two distinct roles, one as the index for the measure and another as center of that measure. Thus, we shall consider for the parameter space a totally arbitrary set Θ and replace the center of the gaussian measure by $m(\theta)$ where $\theta \rightsquigarrow m(\theta)$ is some function defined on Θ. One can always suppose that there is a point

$s \in \Theta$ such that $m(s) = 0$, replacing x by $x - m(s)$ if needed.

With this notation the log likelihood takes the form:

$$\log \frac{dP_\theta}{dP_s} = m(\theta)'Y - \frac{1}{2}\|m(\theta)\|^2$$

where $Y = MX$ is a gaussian random vector.

We shall retain from this remark that the process $\theta \rightsquigarrow \log \frac{dP_\theta}{dP_s}$ is a gaussian process. In view of what precedes, we shall pose the following definition in which Θ is a set without any added structure in which the dimensions are unrestricted, possibly infinite.

Definition 1. *A gaussian experiment indexed by* Θ *is a family* $\mathcal{G} = \{G_\theta; \theta \in \Theta\}$ *of probability measures where*

(1) the G_θ *are mutually absolutely continuous, and*

(2) there is an $s \in \Theta$ *such that the stochastic process of log-likelihood* $\log(dG_\theta/dG_s)$ *is a gaussian process (for the distributions induced by* G_s*).*

Let us temporarily write $X(t) + b(t)$ for the variable $\log(dG_t/dG_s)$, all distributions being taken under G_s, and for $b(t) = E_s \log(dG_t/dG_s)$. The process $t \rightsquigarrow X(t)$ has $E_s X(t) = 0$ and a covariance kernel $K(t,u) = EX(t)X(u)$. We shall see in (iii) below that the notation E without subscript s is justified in the definition of $K(t,u)$ since $K(t,u)$ is the same with respect to any P_θ.

Since $E_s \exp[X(t) + b(t)] = 1$ this implies that the mean $b(t) = -(1/2)E_s X^2(t)$. It is also easy to compute the Hellinger transform of the experiment $\mathcal{G}$. It is given by

$$\begin{aligned}\int \Pi_t (dG_t)^{\alpha(t)} &= E_s \exp\{\sum_t \alpha(t)[X(t) + b(t)]\} \\ &= \exp\{\frac{1}{2}[\sum_t \sum_u \alpha(t)\alpha(u)K(t,u) \\ &\quad - \sum_t \alpha(t)K(t,t)]\}\end{aligned}$$

where $\alpha(t) \geq 0$ and $\sum_t \alpha(t) = 1$, the function α vanishing by assumption except on a finite subset of Θ. In particular the affinity is

$$\begin{aligned} \rho(t,u) &= \int \sqrt{dG_t dG_u} \\ &= \exp\{\frac{1}{8}[2K(t,u) - K(t,t) - K(u,u)]\}. \quad (4.1) \end{aligned}$$

These formulas have several immediate consequences:

(i) The definition of gaussian experiment seems to give a point s a special role. However, if the process $t \rightsquigarrow X(t)$ is gaussian under G_s, it is also gaussian under any other $G_u; u \in \Theta$ since in the Hellinger transform $\int \prod_t (dG_t)^{\alpha(t)}$ the s does not play any special role.

(ii) Call two experiments $\mathcal{E}$ and $\mathcal{F}$ of the same type if they are equivalent in the sense of Chapter 2. The type of a gaussian experiment is entirely determined by the covariance kernel K, since it determines the Hellinger transform. This also means that the type of $\mathcal{G}$ is determined by the affinity function $\rho(t,u) = \int \sqrt{dG_t dG_u}$. Indeed, let $q^2 = -8 \log \rho$; then a bit of algebra shows that

$$K(t,u) = -\frac{1}{2}[q^2(t,u) - q^2(s,t) - q^2(s,u)].$$

(iii) The covariance of the process $t \rightsquigarrow X(t)$ is independent of the measure G_s used to induce the distributions of the process $t \rightsquigarrow X(t)$. If, instead of G_s, one uses G_u, a term $K(t,u)$ is added to the expectation of $X(t)$. This results from a simple computation already carried out in Chapter 3, after the proof of Proposition 1. It also justifies the definition as a gaussian *shift* experiment, since the distributions of the loglikelihood process X are just shifted by changing the value of the parameter. We shall introduce a framework where this is even more visible.

It is now time to introduce the Hilbert space attached to $\mathcal{G}$, and to link our concepts to familiar ones, but first a few words about notation. The alternate formula for the Hellinger transform involves finite sums such as $\sum_t \alpha(t)X(t)$. These would often be written $\sum_j \alpha(t_j)X(t_j); j = 1, 2, ..., k$. This common notation involves an ordering of the finite set where $\alpha(t) \neq 0$. It is a useless complication except for numerical computations. We shall use a different notation, writing sums as integrals. For this, let $\mathcal{M}$ be the space of finite signed measures with finite support on Θ. A finite sum such as the one mentioned above can then be written in the form $\int X d\mu$ or $\int X(t)\mu(dt)$ for a suitable element μ of the vector space $\mathcal{M}$.

We shall also need the linear subspace $\mathcal{M}_0$ of $\mathcal{M}$ formed by those signed measures $\mu \in \mathcal{M}$ such that $\mu(\Theta) = 0$. Its use leads to simpler formulas.

Let $\Lambda(t, s) = \log(dG_t/dG_s)$. Consider the integral $\int \Lambda(t, s)$ $\mu(dt)$. Note that for $\mu \in \mathcal{M}_0$ this sum does not depend on s. Explicitly

$$\int \Lambda(t, s_1)\mu(dt) = \int \Lambda(t, s_2)\mu(dt) \tag{4.2}$$

for any two elements s_1 and s_2 of Θ.

We shall give $\mu \in \mathcal{M}_0$ a norm (or seminorm) $\|\mu\|$ by putting $\|\mu\|^2$ equal to the variance of $\int \Lambda(t, s)\mu(dt)$. This variance is, by remark (iii), independent of the G_s used to induce the distributions. It can be written as

$$\begin{aligned} \|\mu\|^2 &= \int\int K(t, u)\mu(dt)\mu(du) \\ &= -\frac{1}{2}\int\int q^2(t, u)\mu(dt)\mu(du), \end{aligned} \tag{4.3}$$

where $q^2 = -8 \log \rho$ given in (ii). This seminorm is of Hilbertian type. That is, it satisfies the median or parallelogram identity

$$\left\|\frac{\mu + \nu}{2}\right\|^2 + \left\|\frac{\mu - \nu}{2}\right\|^2 = \frac{1}{2}[\|\mu\|^2 + \|\nu\|^2]. \tag{4.4}$$

As a consequence, one can construct a Hilbert space $\mathcal{H}$ attached to the experiment $\mathcal{G}$. To do this, one identifies to zero the $\mu \in \mathcal{M}_0$ such that $\|\mu\| = 0$ and one completes the resulting space.

It is possible to proceed somewhat differently and at the same time extend the parameter set Θ of $\mathcal{G}$ to the whole space $\mathcal{H}$. Take the particular $s \in \Theta$ used before. If $\theta \in \Theta$, represent it by the measure $\delta_\theta - \delta_s$, which gives mass unity to θ and mass (-1) to s. Using the centered loglikelihood process $X(t) = \Lambda(t,s) - E_s\Lambda(t,s)$, form the integrals $\langle X, \mu\rangle = \int X(t)\mu(dt)$, with $\mu \in \mathcal{M}_0$. These are gaussian random variables. The map $\mu \rightsquigarrow \langle X, \mu\rangle$ is a continuous linear map from $\mathcal{M}_0$ to the space of gaussian variables, if one metrizes that space by the convergence in quadratic mean. The map extends by continuity in quadratic mean to the space $\mathcal{H}$. For the elements of $\mathcal{H}$ we shall use the same notation as for $\mathcal{M}_0$.

Note that $E_s\langle X, \mu\rangle = 0$ and $E_s[\langle X, \mu\rangle]^2 = \|\mu\|^2$. Thus there is a probability measure $G_{(\mu)}$ such that

$$dG_{(\mu)} = \exp[\langle X, \mu\rangle - \frac{1}{2}\|\mu\|^2]dG_s. \tag{4.5}$$

According to this construction G_θ becomes $G_{(\delta_\theta - \delta_s)}$. This notation is a bit cumbruous. Thus, we shall drop the parenthesis around the subscripts such as (μ) and identify the point θ of Θ to the element $\delta_\theta - \delta_s$ of $\mathcal{H}$. This should not cause any major difficulty.

There is no difficulty computing various entities related to our new gaussian measures. For example, the Hellinger affinity between G_μ and G_ν becomes

$$\begin{aligned}\int \sqrt{dG_\mu dG_\nu} &= \int \exp[\frac{1}{2}\langle x, \mu + \nu\rangle - \frac{1}{4}(\|\mu\|^2 + \|\nu\|^2)]dG_s \\ &= \exp\{-\frac{1}{8}\|\mu - \nu\|^2\}.\end{aligned}$$

This is easily derivable by evaluating $E_s \exp\{\frac{1}{2}\langle X, \mu - \nu\rangle\}$.

The function q used in equation (4.3) above has a simple meaning. By (4.3) one has $q(t,u) = \|\delta_t - \delta_u\|$. As such it is a "natural distance" on the set Θ. There is a slight difficulty that might require elaborate circumlocutions: *There may be distinct points t and u such that $q(t,u) = 0$. To avoid that problem we shall assume that $q(t,u) = 0$ implies that t and u are the same.* In other words, we shall assume that q is a metric on Θ.

Now consider not one but two gaussian experiments $\mathcal{G}_i; i = 1, 2$. They correspond to sets Θ_i and distances q_i. Let us assume that there is a one-to-one map, say, j, that is an isometry of Θ_1 onto Θ_2; that is,

$$q_2(j(t), j(u)) = q_1(t, u) \quad \text{for all pairs } (t, u).$$

If s has been taken as the origin in Θ_1, take $j(s)$ as the origin in Θ_2. Let $\mathcal{H}_i$ be the corresponding Hilbert spaces.

Then the isometry of the Θ_i extends to a linear isometry between the $\mathcal{H}_i$.

(If we had not matched the origins, it would be an affine isometry.) This statement is a simple consequence of how the $\mathcal{H}_i$ were constructed. One implication is that the experiments $\mathcal{G}_1$ and $\{G_{j(t)}; t \in \Theta_1\}$ are equivalent. In other words, the *statistical* properties of a gaussian experiment, as reflected in the available risk functions for arbitrary bounded loss functions, depend only on the *geometrical* properties of its parameter set in the attached Hilbert space.

In the preceding discussion we have focused mostly on what happens to the parameter space and its extension. But if one attaches to $\mu \in \mathcal{M}_0$ the random variable $\int X(t)\mu(dt)$ one obtains a random process that extends by continuity to $\mathcal{H}$ an example of what has been called the *canonical* or the *isonormal* gaussian process of the Hilbert space $\mathcal{H}$. Further discussion is given in example (b). See also Dudley [1989] for its properties.

The gaussian experiment itself is obtained by a set of shifts of the isonormal process of $\mathcal{H}$; the shifts are also elements of $\mathcal{H}$. This being said, it is time to look at some examples.

Examples.

(a) Our first example is the finite-dimensional gaussian shift introduced at the beginning of this section. One has a k-dimensional random column vector X. It is gaussian with mean zero and covariance matrix $C = EXX'$. What is observed is a vector $Y = \theta + X$. The center θ is allowed to roam over the entire k-dimensional space $\Re^k$. Here

$\Lambda(t, 0) = X(t) - (1/2)t'Mt$ where $X(t)$ is simply the inner product $t'MX$ and M is the inverse of the covariance matrix C.

The integrals with respect to a $\mu \in \mathcal{M}_0$ can be written:

$$\int X(t)\mu(dt) = [\int t\mu(dt)]'MX.$$

That is, the map $t \rightsquigarrow \int t\mu(dt)$ maps $\mathcal{M}_0$ linearly onto $\Re^k$ and the seminorm is given by $\|\mu\|^2 = [\int t\mu(dt)]'M[\int t\mu(dt)]$. The resulting Hilbert space $\mathcal{H}$ is isomorphic to $\Re^k$ with the square norm $\theta' M \theta$.

Note that if C was singular, some of the corresponding measures G_t and G_u would be singular. Definition 1 does not allow that possibility. One could consider the case where M is singular. This can happen easily when taking limits of experiments. Then our Hilbert space $\mathcal{H}$ is no longer isomorphic to $\Re^k$ but to a quotient of it, identifying to zero those t such that $t'Mt = 0$.

This example points out a general fact. If the parameter set Θ is a linear space with a centered $\Lambda(t, 0)$ linear in $t \in \Theta$, if $q(t, u) = 0$ implies $t = u$, and if Θ is complete for the distance q, then $\mathcal{H}$ coincides with Θ metrized by q. If any of the stated conditions is violated, one may need to think further and construct the appropriate $\mathcal{H}$.

(b) A more interesting example is given by what is called the "signal plus noise model." In its simplest form it looks like this: One has an interval of time, say, $[0, 1]$. The time will be denoted by τ to avoid confusion with our previous use of t as a parameter for the experiment. One observes a process $Y(\tau)$ such that $dY(\tau) = g(\tau)d\tau + \sigma dW(\tau)$ where W is a standard Wiener process so that dW is "white gaussian noise." Here the parameter of the experiment is the function $\tau \rightsquigarrow g(\tau)$. For the reader who may be distracted by the differential notation, note that it may be interpreted by integration as $Y(\tau) = \int_0^\tau g(\xi)d\xi + \sigma W(\tau)$ or by integration of smooth functions as in the case of L. Schwartz's distributions. The names "white noise" and "Wiener process" need some elaboration, which can be given in terms of the *canonical* or *isonormal* gaussian process of a Hilbert space. (we have said 'the' in spite of the fact that there can be many such processes, but they are equivalent in distribution.)

Let $\mathcal{X}$ be a Hilbert space with the norm of x denoted $\|x\|$. An isonormal or canonical gaussian process of $\mathcal{X}$ is a map $x \rightsquigarrow Z(x)$ where $Z(x)$ is a gaussian random variable such that $EZ(x) = 0$ and $E|Z(x)|^2 = \|x\|^2$. The map is in addition assumed to be linear. That such maps exist can be seen in various ways. One way is to note that it would lead to a characteristic function $E\exp[iZ(x)] = \exp\{-\|x\|^2/2\}$. This function is of positive type. Thus, according to a theorem of Bochner [1947], there is on the algebraic dual of $\mathcal{X}$ a probability measure that generates the desired map.

One can proceed differently and take an orthonormal basis $\{u_\alpha\}$ in $\mathcal{X}$. Then one provides oneself with a family ξ_α of mutually independent gaussian variables, each with mean zero and variance unity. If $x = \sum_\alpha b(\alpha)u_\alpha$, let $Z(x) = \sum_\alpha b(\alpha)\xi_\alpha$. (Infinite sums are defined by convergence in quadratic mean.)

This type of construction occurs often. To get the white noise, let λ be the Lebesgue measure. Let $\mathcal{X}$ be the Hilbert space of equivalence classes of λ-square integrable functions, with the square norm $\|g\|^2 = \int g^2 d\lambda$. Construct the canonical process Z attached to that Hilbert space. To obtain W, one takes a separable version of $\tau \rightsquigarrow W(\tau) = Z(I_\tau)$, where I_τ is the indicator of the interval $[0, \tau)$.

With all of these explanations the model $dY(\tau) = g(\tau)d\tau + \sigma dW(\tau)$ makes sense. To get a gaussian experiment $\{G_\theta\}$ one assumes that the coefficient σ is *known*, and one takes a set Θ of functions g. To assure absolute continuity of the G_θ, one assumes that if $g_i \in \Theta$, then $\int [g_1 - g_2]^2 d\lambda < \infty$. This is readily seen to be equivalent to the fact that

Hellinger affinities do not vanish.

Let us assume, for simplicity, that the function g identically zero belongs to Θ. For square norm, use $\sigma^{-2} \int g^2 d\lambda$. Then our Hilbert space $\mathcal{H}$ is the closure of the linear space spanned by the set of equivalence classes of the $g \in \Theta$ in the space L_2 of equivalence classes of square integrable functions. Note that $\mathcal{H}$ may be finite-dimensional, as in the use of limited Fourier expansions. However, $\mathcal{H}$ may be infinite-dimensional even if the g's belong to a "parametric" family in the (imprecise) sense that they depend smoothly on some euclidean parameter. Consider, for example, the functions $g_\theta = \exp(\theta\tau)$ with θ real or complex.

One could worry about using σ as part of the parametrization. This would not work here since the distributions of processes that correspond to different σ are disjoint, as follows, for example, from computations of affinities.

Let us conclude the description of this example with two remarks. The first has to do with the fact, noted earlier, that a gaussian experiment is defined up to equivalence by the *geometric* properties of its index set as embedded in Hilbert space. This implies that any gaussian experiment $\mathcal{G}$ such that the parameter space Θ is metrized and separable for the attached function q can be realized by a scheme $dY(\tau) = g(\tau)d\tau + dW(\tau)$ as above (with $\sigma = 1$).

The second remark is that Nussbaum [1996] has shown that, under weak conditions on the set Θ of densities f and for n i.i.d. observations, the experiment is approximated in the sense of our distance (Chapter 2, Definition 2) by one corresponding to $dY = gd\tau + \sigma dW$ where $g = \sqrt{f}$ and $\sigma = 1/2\sqrt{n}$. This result has many important implications for nonparametric or semiparametric studies, since it allows treating such problems approximately "as if" they were gaussian shift ones.

(c) Consider a space V in which a finite set $\{v_j; j \in J\}$ is selected arbitrarily. For each j one observes a random variable Y_j. Assume that Y_j is gaussian with expectation $g(v_j)$ and standard deviation $\sigma(v_j)$. This is a fairly standard regression problem. Here we shall assume that the function $v \rightsquigarrow \sigma(v)$ is known and that the "parameter" is the function g. To put this in the general framework, define a distance by a norm $\|g\|$ where

$$\|g\|^2 = \sum_j \frac{g(v_j)^2}{\sigma^2(v_j)}$$

as one does for the method of least squares. One selects a set Θ of functions g as the parameter space. This is often a linear space spanned by some finite set $\{g_i\}$ of functions. It can be a set of splines with prescribed degree and a selected number of nodes. The position of each node is specified. The case of nodes of unspecified position but restricted in number is a bit more complicated. In any event, our Hilbert space $\mathcal{H}$ is a finite-dimensional space . It takes into account only the values $g(v_j)$, with the "least squares" norm, for which it is automatically complete, being finite-dimensional.

Here a problem of different nature can surface. Suppose that a particular set Θ of functions g defined on V has been specified. Suppose also that the statistician is allowed the *choice* of the finite set $\{v_j; j \in J\}$, as occurs in the design of experiments. Since it is assumed that the function $\sigma : v \rightsquigarrow \sigma(v)$ is known, this gives the statistician a choice of distances on Θ, each choice of set $\{v_j\}$ corresponding to a choice of distance and therefore of norm on the (finite-dimensional) Hilbert space. The experiments corresponding to two distances may (or may not) be comparable. They may be close to each other (or may not) for our distance Δ between experiments.

(d) The following type of example shows that, when embedded in the associated Hilbert space $\mathcal{H}$, the parameter set Θ can be a very "thin" set, although of infinite *linear* dimension.

To explain, call a square distance d^2 of *negative type* if for every $\mu \in \mathcal{M}_0$ one has

$$\int\int d^2(s,t)\mu(ds)\mu(dt) \leq 0.$$

There are numerous examples of such distances. For example, if Θ is the real line, one can take $d(s,t) = |s-t|^\alpha$ for some $0 < \alpha < 1$. One can define a seminorm exactly as was done for q in (4.3) and construct the associated Hilbert space $\mathcal{H}_d$. To this Hilbert space is attached a gaussian experiment, parametrized by $\mathcal{H}_d$ itself. One can obtain it by considering the shifts by elements of $\mathcal{H}_d$ of the isonormal process of $\mathcal{H}_d$. If this is carried out for the distance $d(s,t) = |s-t|^\alpha$ on the line and if one takes for Θ the line itself, the image of Θ in $\mathcal{H}_d$ is a curve, more specifically a helix of topological dimension unity that spans linearly the infinite-dimensional $\mathcal{H}_d$.

For $\alpha = 1/2$, the Hilbert space associated with d is the space of equivalence classes of Lebesgue square integrable functions. The corresponding gaussian process is the white noise on the line. This is as in Example *(b)*, but without any restriction on the range of τ. For $0 < \alpha < 1/2$, the corresponding integrated gaussian process is often called " fractional brownian motion." It is not directly related to brownian motion itself.

Among the usual examples, one should also mention the cases, as in the log normal distributions, where the observations are one-to-one functional transforms of gaussian ones. This would cover the so-called "log normal" where the observable variables Y are of the form $Y = \exp\{aX+b\}$, with a known and with X gaussian with known variance, but definitely not the "three-parameter" case where the observables are of the form $Y = c+\exp\{aX+b\}$ and where a shift by c gives them different supports.

We now need to say a few words about the statistical aspects.

In the general formulation, the parameter $\mu \in \mathcal{H}$ appears as a shift parameter. That might sound strange but one should remember that μ appeared in $\mathcal{H}$ as a result of multiple avatars. To see that it is so, note that under G_0 the process $\{\langle X, \nu\rangle; \nu \in \mathcal{H}\}$ has expectation zero. Under G_μ its expectation becomes $\langle \nu, \mu\rangle$ for an inner product obtained from the Hilbert norm $\|\mu\|$ as usual. In addition, the variances are not changed. Thus, the process $\langle X, \nu\rangle$ is just shifted by $\langle \nu, \mu\rangle$.

Another way of expressing this can be used if $\mathcal{H}$ is finite-dimensional. Our notation $\langle X, \mu\rangle$ suggests that X should be regarded as a random vector with values in $\mathcal{H}$ itself. This is legal if one identifies $\mathcal{H}$ with its dual. One can always perform such an identification. If f is a continuous linear functional defined on $\mathcal{H}$, there is an element $\nu_f \in \mathcal{H}$ such that $f(\mu) = \langle \nu_f, \mu\rangle$ identically. (Note that this identification depends on the norm $\|.\|$ used.) Considering X as random with values in $\mathcal{H}$ is pleasant, but when $\mathcal{H}$ is infinite-dimensional, our X does not come from a true "distribution" on $\mathcal{H}$, only a "cylinder measure," since $\|X\| = +\infty$ almost surely. However, in the finite-dimensional case, one may write

$$\begin{aligned} \log \frac{dG_\mu}{dG_0} &= \langle X, \mu\rangle - \frac{1}{2}\|\mu\|^2 \\ &= -\frac{1}{2}[\|X - \mu\|^2 - \|X\|^2]. \end{aligned}$$

From this, assuming that the parameter set is the whole $\mathcal{H}$, it follows that:

1. The observed X is the maximum likelihood estimate of μ.

2. If T is an equivariant estimate of μ, that is, if $T(X+h) = T(X) + h$ for all $h \in \mathcal{H}$, then the distribution of T is a convolution of the gaussian distribution of X with some other distribution. In particular, X is "more concentrated" than T.

3. For a large variety of loss functions, X is minimax.

4. Unfortunately, as shown by C. Stein, L.D. Brown and others, if the dimension of $\mathcal{H}$ exceeds 2, then X is not an admissible estimate of μ for a very large class of loss functions

such that the loss $W(T, \theta)$ depends only on the difference $T - \theta$, identifying θ to its image in $\mathcal{H}$ if necessary. Typical examples would be $W(t, \theta) = \|t - \theta\|^2$ or, taking a coordinate system, $\sup |T_i - \theta_i|$. If the dimension of Θ is at all large, possible improvements can be substantial.

These properties remain true in infinite-dimensional cases, as occurs in the case of density estimation or other nonparametric situations, but one must interpret the word "distribution" as "cylinder measure." The form of loglikelihood ratio $-[\|X - \mu\|^2 - \|X\|^2]/2$ is no longer available because the norms are infinite. The first equality stays.

The lack of admissibility of X as an estimate does not reduce its value as a "statistic." Indeed it is sufficient and, in many cases, minimal sufficient. It does, however, call for precautions in estimation or testing. Some remedies have been proposed in the form of the shrinkage estimates of C. Stein, the "ridge regression" procedures or the hard and soft thresholding methods of D. Donoho. We do not have the space to discuss these methods here. In addition, it will be shown that in "parametric" asymptotics, improvements are feasible only on very "small" sets; see Chapter 6.

Finally, note that X can hardly ever be the Bayes estimate for a proper prior probability measure. (One could doctor the loss functions to make it a proper Bayes estimate, at least in some usual situations.)

In view of these improvements of C. Stein and L. Brown, even in the gaussian shift case, the reader should ponder the statements attributed to R. A. Fisher about the superiority of the maximum likelihood method or some statements by various authors, e.g., Lehmann, *Theory of Point Estimation* (page 170), to the effect that the "principle of equivariance" is very convincing. The principle simplifies the mathematical derivations, but this may be at the expense of a sizable increase in the risk functions. In this respect, it is not more justifiable than the "principle of unbiasedness."

4.3 Poisson Experiments

Let $\mathcal{X}$ be a set and let $\mathcal{A}$ be a ring of subsets of $\mathcal{X}$. A *content* λ on $\mathcal{A}$ is a finite set function $A \rightsquigarrow \lambda(A)$ from $\mathcal{A}$ to $[0, \infty)$ such that λ is additive: If A and B are disjoint then $\lambda(A \cup B) = \lambda(A) + \lambda(B)$. A Poisson process of base λ is a function Z to random variables such that:

1. If A and B are disjoint then $Z(A \cup B) = Z(A) + Z(B)$.
2. If A_i are pairwise disjoint, then the $Z(A_i)$ are independent.
3. The variable $Z(A)$ is Poisson with expectation $\lambda(A)$.

As explained in the introduction to this chapter, we use the word "content" because there are too many things called "measures" floating around and because we shall not perform infinite operations on $\mathcal{A}$, but the reader may want to assume that λ is a measure.

According to this definition Z is a random function of *sets*. It is often more convenient to regard Z as a random linear functional. To do this, introduce the space $\mathcal{S}$ of simple functions $v = \sum a_j I(A_j)$ where the a_j are numbers and $I(A_j)$ is the indicator of an element $A_j \in \mathcal{A}$, and the A_j are disjoint. The sum is supposed to be a finite sum. One will then write

$$\int v dZ = \int v(x) Z(dx) = \sum_j a_j Z(A_j). \tag{4.6}$$

To be complete, one should show that, given a content λ, there is a Poisson process Z with $EZ(A) \equiv \lambda(A) : A \in \mathcal{A}$, and such a process has a uniquely defined distribution. This can be seen as follows.

First take a set $C \in \mathcal{A}$ and let $\{A_j\}$ be a partition of it by elements of $\mathcal{A}$. Let N be a Poisson variable with expectation $\lambda(C)$. Let $\pi_j = \lambda(A_j)/\lambda(C)$. If N takes value n, distribute n points at random in C so that the probability of falling in A_j is π_j. This will give a Poisson process on the ring generated by the chosen partition. It will correspond to the characteristic function

$$\phi(v) = E[\exp\{i \int v dZ\}] = \exp\{\int (e^{iv} - 1) d\lambda\}. \tag{4.7}$$

To go further, one can appeal to Bochner's theorem if one shows that the function ϕ is of positive type. However, note that the property of being of positive type involves only a finite set of functions $v \in \mathcal{S}$ at a time. Thus, one can reduce oneself to the special case of finite partitions just considered.

This proves the existence of the process Z. The uniqueness in distribution follows from the fact that $\phi(v)$ is well determined for a v defined by a finite partition.

Definition 2. *A Poisson experiment $\{P_\theta; \theta \in \Theta\}$ indexed by Θ is given by a ring $\mathcal{A}$ of subsets of a set $\mathcal{X}$ and a map $\theta \rightsquigarrow \mu_\theta$ from Θ to contents on $\mathcal{A}$. The probability measure corresponding to θ is the distribution P_{μ_θ} of the Poisson process Z of base μ_θ, which will be abbreviated by P_θ.*

Note that we took for $\mathcal{A}$ a *ring* not a *field*. This is to avoid infinities, as would happen, for instance, on the real line, taking for λ the Lebesgue measure on the field generated by compact intervals. By contrast, the sets of the ring they generate have finite Lebesgue measure.

Poisson experiments occur in great abundance in modeling natural phenomena, in physical or biological settings or otherwise. One can mention the following:

(a) The most classical example is that of telephone calls, at a private line or at a dispatching center. Here the content λ may be taken to be a multiple of Lebesgue measure, but in more refined studies it may reflect diurnal or seasonal variation. Here $Z(A)$ is the number of calls over a period of time A.

(b) Poisson processes have also been used to describe the arrival of claims together with their amounts at an insurance company. The space we called $\mathcal{X}$ is then a product of two lines, one for the time and the other for the financial amount. The value of $Z(A \times B)$ is then the number of claims that occurred in A and whose value was in the range B.

(c) Hemacytometer counts of lymphocytes or red blood cells are said to lead to Poisson processes. This is also the case for bacterial counts in drinking water, for instance, in lakes or reservoirs. The federal regulations on these subjects assume the applicability of Poisson processes. This is also true for various particulate counts, even sometimes when the Poisson model is demonstrably inapplicable, as in the case of asbestos fibers in the air. (The fibers are electrically charged and repel one another.)

(d) Physicists often assume that the lifetime of a radioactive atom is governed by an exponential distribution, $Pr[T > x] = e^{-\lambda x}$. This leads to a Poisson process for the number of disintegrations over a time interval in a batch of the radioactive substance. Of course, the phenomenon of chain reactions that take place in an atom bomb are not included.

(e) Seismologists have attempted to describe the place and time of occurrence of earthquakes by Poisson processes. Here the process is in a four-dimensional space, where three dimensions record the position of the epicenter and the fourth records the initial time of occurrence, excluding aftershocks.

(f) The position and time of occurrence of car accidents on roadways have also been modeled through Poisson processes. Here a geographical estimate of the intensity of car accidents is important. It may lead to improved signalization, lighting, or pavement.

This is far from a complete list, but it should be sufficient to indicate the practical relevance of Poisson experiments. In addition, the mathematical importance of Poisson experiments will be stressed in connection with infinite divisibility. Let us start with some basic formulas. For these, we shall consider a Poisson experiment $\mathcal{E}$ according to Definition 2. Recall that the Hellinger transform of an experiment is a function of α defined by $\alpha = \{\alpha_\theta; \theta \in \Theta\}$ such that $\alpha_\theta \geq 0$, $\sum \alpha_\theta = 1$ but with only a finite set of α_θ different from zero.

Proposition 1. *The Hellinger transform of $\mathcal{E}$ is given by*

$$\phi(\alpha) = \int \Pi_\theta (dP_\theta)^{\alpha_\theta} = \exp\left\{\int \left[\Pi_\theta (d\mu_\theta)^{\alpha_\theta} - \sum \alpha_\theta d\mu_\theta\right]\right\}.$$

In particular the affinity between two probability measures P_s, P_t is given by

$$\int \sqrt{dP_s dP_t} = \exp\left\{-\frac{1}{2}\int (\sqrt{d\mu_s} - \sqrt{d\mu_t})^2\right\}. \tag{4.8}$$

Proof. The proposition can be proved by looking first at a family $\{A_j\}$ of pairwise disjoint sets for which the Hellinger transform is given by the product of Hellinger transform for ordinary Poisson distributions on the line. Explicitly, a set A yields a Poisson variable with expectation $\lambda_\theta = \mu_\theta(A)$ and probabilities $p_\theta(k) = e^{-\lambda_\theta} {\lambda_\theta}^k / k!$. This gives a Hellinger transform

$$\phi_A(\alpha) = \exp\{-\sum \alpha_\theta \lambda_\theta + \prod_\theta \lambda_\theta^{\alpha_\theta}\}.$$

Each A_j gives such a transform. Since the Poisson variables attached to the A_j are independent, the Hellinger transform obtained from the simultaneous observation of the $Z(A_j)$ is the product over j of such expressions. That product has exactly the form given in the proposition, except that the integral sign is replaced by a sum over j and $d\mu_\theta$ is replaced by $\mu_\theta(A_j)$. To obtain the form given in the proposition one carries out two limiting procedures: One refines indefinitely the partition of the union $B = \bigcup_j A_j$ and one lets this set B expand to the whole space. The procedure is the classical one carried out by Hellinger [1909] to define his integral. It is also described in Bochner [1955] in connection with the characteristic functionals of additive random set functions. □

One of the consequences of the formula for the Hellinger affinity is that one can partition the set Θ as $\Theta = \bigcup_j \Theta_j$ so that if μ and ν correspond to the same Θ_j then $\int(\sqrt{d\mu} - \sqrt{d\nu})^2 < \infty$, while if they come from different sets, say, Θ_j and Θ_k, the integral is infinite and the corresponding Poisson measures disjoint. We shall refer to this situation by saying that the Poisson experiment corresponding to one set Θ_j is *pairwise imperfect.* This does not mean that the measures constituting it are mutually absolutely continuous, just that they are not pairwise disjoint. There is no equivalent here to the Kakutani alternative or the Feldman-Hájek theorem according to which gaussian measures are either disjoint or mutually absolutely continuous. Consider two contents μ and ν on the same ring and let P_μ be the distribution of the Poisson process of intensity μ; similarly for P_ν. Following the same argument and the same computation as in Section 3.2, one sees that if $\int(\sqrt{d}\mu - \sqrt{d}\nu)^2 < \infty$, then the mass of P_μ that is singular with respect to P_ν is given by $1 - \exp[-\beta]$ where β is the mass of μ that is ν singular.

The formula also entails that for a pairwise imperfect Poisson experiment there exists a gaussian experiment with exactly the same affinities. This shows that even though the Hellinger tranforms uniquely determine the type of an experiment, affinities are not enough to do the same. They are enough if one specifies that the *experiment is gaussian,* as was shown in Section 4.2. However, if one replaces the term "gaussian" by "Poisson", the uniqueness no longer holds.

Now let us pass to some considerations that underlie the appearance of Poisson experiments in many instances. Any experiment, say, $\mathcal{E}$, has a *Poissonized* version, say, $ps\mathcal{E}$ obtained as follows. One first observes an ordinary Poisson variable N independent of anything else such that $EN = 1$. If N takes the value k, one carries out k independent replicates of $\mathcal{E}$. If one has a set $\mathcal{E}_j; j \in J$ of experiments carried out independent of each other, one associates to the direct product $\bigotimes \mathcal{E}_j$ the *accompa-*

nying infinitely divisible experiment $\bigotimes ps\mathcal{E}_j$. This accompanying experiment is itself a Poisson experiment, as can easily be verified as follows. If the experiment $\mathcal{E}_j$ has Hellinger transform $\phi_j(\alpha)$, that of $ps\mathcal{E}_j$ is

$$\psi_j(\alpha) = \exp(\phi_j(\alpha) - 1).$$

Taking direct products of experiments is represented by taking the pointwise product of their Hellinger transforms; see Chapter 3. For these one can write inequalities as follows. The α in the argument is arbitrarily fixed and hence omitted. Let

$$\gamma_1 = \frac{1}{2} \sup_j [1 - \phi_j].$$

If the set J has finite cardinality equal to m, introduce the average $\bar{\phi} = \sum_j \phi_j / m$ and the expression

$$\gamma_2 = \frac{1}{m} \sum_j [\phi_j - \bar{\phi}]^2 [1 - \bar{\phi}]^{-2}.$$

If $\bar{\phi} = 1$ for a value of α, interpret the value of γ_2 as zero.

Proposition 2. *Let* $\phi = \Pi_j \phi_j$ *and let* $\psi = \exp\{\sum_j(\phi_j - 1)\}$. *Then for every* α,

$$0 \le \psi - \phi \le e\gamma$$

where γ *is the minimum of* γ_1 *and* $[1 + \gamma_2]/m$.

Proof. It is enough to prove the results for the case where J is the segment $\{1, 2, \ldots, m\}$ of the integers, extensions to other possible sets J being immediate. Note then the inequalities $0 \le \phi_j \le \psi_j \le 1$. They imply, in particular, that

$$0 \le \psi_j - \phi_i = e^{\phi_j - 1} - \phi_j \le \frac{1}{2}(1 - \phi_j)^2.$$

Introduce partial products

$$C_j = (\Pi_{k<j} \psi_k)(\Pi_{i>j} \phi_i).$$

Then one has $\psi - \phi = \sum_j C_j(\psi_j - \phi_j)$. Since $C_j \leq \psi\psi_j{}^{-1}$ and $\psi_j \geq e^{-1}$; this gives

$$0 \leq \psi - \phi \leq \frac{e\psi}{2} \sum_j (1 - \phi_j)^2.$$

Put

$$\omega = \sum_j (1 - \phi_j)$$

so that $\psi = e^{-\omega}$ and note that $\omega e^{-\omega} \leq 1$. Since

$$\sum (1 - \phi_j)^2 \leq \omega \sup_j (1 - \phi_j) = \omega 2\gamma_1,$$

this yields the first inequality.

For the second, note that

$$\begin{aligned} \sum (1 - \phi_j)^2 &= \sum (\bar{\phi} - \phi_j)^2 + \sum (1 - \bar{\phi})^2 \\ &= (1 + \gamma_2) m (1 - \bar{\phi})^2 \\ &= (1 + \gamma_2) m^{-1} \omega^2. \end{aligned}$$

Since $\omega^2 e^{-\omega} \leq 2$, this yields the second inequality. □

The reader should be reminded that these computations involve Hellinger transforms, which are functions of α. As a result, the γ_i are also functions of this same α, even though it was suppressed to simplify the formulas.

The inequality involving γ_1 was meant for cases where each of the experiments $\mathcal{E}_j$ viewed separately contains "little information." The inequality involving γ_2 was intended for the case where the $\mathcal{E}_j$ are replicates of the same experiment or close to such replicates.

However, the relations between the general distance Δ of Chapter 2 and, say, uniform distance (over α) between Hellinger transforms are not as simple as one would hope. So we shall consider only a few simple implications. The first few involve sets Θ that are finite, of cardinality k.

Proposition 3. *There is a function $\epsilon \leadsto b_k(\epsilon)$ tending to zero as ϵ tends to zero, with the following property:*

Let $\mathcal{E}_i; i = 1, 2$ be two experiments indexed by a set Θ of finite cardinality $\text{card}\Theta \leq k$. Let ϕ_i be their respective Hellinger transforms. If

$$\sup_{\alpha} |\phi_1(\alpha) - \phi_2(\alpha)| \leq \epsilon$$

then

$$\Delta(\mathcal{E}_1, \mathcal{E}_2) \leq b_k(\epsilon).$$

Proof. Assuming the contrary, one could find sequences $\{\mathcal{E}_{i,n}\}$ and a number $b > 0$ such that the differences $\phi_{1,n} - \phi_{2,n}$ tend to zero uniformly but $\Delta(\mathcal{E}_{1,n}, \mathcal{E}_{2,n}) > b$. The compactness of the set of experiment types indexed by Θ (see Chapter 2, Lemma 4) allows us to assume that the $\mathcal{E}_{i,n}$ tend to limits, say, $\mathcal{F}_i$. The limits of the Hellinger transforms are those of the $\mathcal{F}_i$, by Chapter 3. Thus they must be the same. However, since Hellinger transforms determine the experiments, that contradicts the implied inequality $\Delta(\mathcal{F}_1, \mathcal{F}_2) \geq b$. □

A reverse implication also holds, but only pointwise in α. There is still a special case of interest, as follows. Recall that a trivial experiment is the one in which all probability measures are the same and thus its Hellinger transform is equal to 1.

Proposition 4. *Let Θ have cardinality k. Let $\mathcal{T}$ be the trivial experiment. Let $\mathcal{E}$ be an experiment indexed by Θ such that $\Delta(\mathcal{E}, \mathcal{T}) \leq \epsilon$. Then the difference between the Hellinger transform ϕ of $\mathcal{E}$ and that of the trivial experiment satisfies*

$$0 \leq 1 - \phi \leq 4k\epsilon.$$

Proof. Pretend that the set of indices Θ is the ordered set of integers $\{1, 2, \ldots, k\}$ so that $\mathcal{E} = \{P_j, j = 1, \ldots, k\}$. The inequality $\Delta(\mathcal{E}, \mathcal{T}) \leq \epsilon$ implies the existence of a measure that is in L_1-norm within 2ϵ of each of the P_j; see Chapter 2. In particular if $\alpha = \{\alpha_j, j = 1, 2, \ldots, k\}$ is a given argument for ϕ and $Q = \sum \alpha_j P_j$, one has $\|Q - P_j\| \leq 4\epsilon$ for all j.

Changing the measures P_j to Q one at a time, one sees that

$$\begin{aligned} 1 - \phi(\alpha) &= \int \Pi(dQ)^{\alpha_j} - \Pi(dP_j)^{\alpha_j} \\ &= \sum_{j=1}^{k} \int (dM_j)^{1-\alpha_j}[(dQ)^{\alpha_j} - (dP_j)^{\alpha_j}] \end{aligned}$$

where M_j is the geometric mean

$$(dM_j)^{1-\alpha_j} = \Pi_{i<j}(dQ)^{\alpha_j} \, \Pi_{r>j}(dP_j)^{\alpha_r}.$$

To bound $1 - \phi$ we proceed as follows. Since the geometric mean is no larger than the arithmetic mean, one has

$$(dM_j)^{1-\alpha_j}(dQ)^{\alpha_j} \leq 2dQ.$$

Take Radon Nikodym densities with respect to Q, getting $f_j = dP_j/dQ$. An upper bound to the jth term in the integral is therefore given by

$$2\int [1 - f_j{}^{\alpha_j}]dQ.$$

Since one has $[1 - f_j{}^{\alpha_j}]^+ \leq [1 - f_j]^+$, the result follows. □

Leaving this aside, let us return to the inequalities of Propositions 2 and 3. Consider an experiment $\mathcal{E} = \bigotimes \mathcal{E}_j$ a direct product of experiments $\mathcal{E}_j$, all indexed by the same set Θ. For a pair (s, t) of elements of Θ define a distance $H(s, t)$ by

$$H^2(s, t) = \sum_j h_j{}^2(s, t),$$

where h_j is the Hellinger distance defined by $\mathcal{E}_j$.

Proposition 5. *Assume that there is a set F of cardinality m such that each element $\theta \in \Theta$ is within H-distance at most ϵ of F. Let $\mathcal{F} = \bigotimes[ps\mathcal{E}_j]$ be the accompanying Poisson experiment. Let $\phi_{\mathcal{E}}$ be the Hellinger transform of $\mathcal{E}$ and similarly for $\mathcal{F}$. Suppose that if $\alpha = \alpha_\theta; \theta \in F$ then $|\phi_{\mathcal{E}} - \phi_{\mathcal{F}}| \leq \epsilon$. Then*

$$\Delta(\mathcal{E}, \mathcal{F}) \leq b_m(\epsilon) + 4\epsilon.$$

This result may seem to be of limited applicability, yet it is in its own way more general than the results to be presented in the next chapter where, in addition to the negibility of the individual experiments, one imposes bounds on their products. A more interesting result is available when the $\mathcal{E}_j$ are all replicates of the same experiment, say, $\mathcal{E}_0$. Taking an integer n and observing an independent Poisson variable N with $EN = n$, one has the experiment $\mathcal{E}$ where $\mathcal{E}_0$ is repeated n times independently to compare with $\mathcal{F}$ where $\mathcal{E}_0$ is repeated N times, also independently.

The inequality given below depends on the possibility of estimating θ from $\min(n, N)$ observations. Since we have not mentioned this so far, the proof will not be given here. See Le Cam [1986, p. 508], where the result is proved but misstated.

The inequality depends on a concept of metric dimension as follows. Metrize Θ by the distance H as in the previous proposition.

Definition 3. *The metric dimension $D(\tau)$ of Θ at the level τ is the smallest number such that every set of H diameter $2x \geq 2\tau$ can be covered by at most $2^{D(\tau)}$ sets of diameter x.*

Theorem 1. *Let $\mathcal{E}$ be the direct product of n independent replicates of an experiment $\mathcal{E}_0$. Let $\mathcal{F}$ be the Poisson experiment direct product of n replicates of $ps\mathcal{E}_0$. Then*

$$\Delta(\mathcal{E}, \mathcal{F}) \leq \frac{10.6}{n^{\frac{1}{4}}} \sqrt{1 + D(\tau)} \tag{4.9}$$

where τ is roughly such that $\tau^2 = CD(\tau)$ for a certain universal coefficient C.

One of the consequences of the preceding propositions is that every infinitely divisible experiment $\mathcal{E}$ is a limit of Poisson experiments. Indeed, for every integer n one can write $\mathcal{E}$ as a direct product of n replicates of some experiment $\mathcal{E}_{0,n}$. The *weak* convergence to $\mathcal{E}$ follows then from Propositions 3 and 4. A stronger form of convergence follows from Theorem 1. Introduce for each

n a dimension function $D_n(\tau)$ as defined by the $\mathcal{E}_{0,n}$. Then the convergence holds in the sense of our distance Δ on any subset of Θ such that $\sup_n D(\tau) < \infty$. In fact one can say more, taking sets such that $n^{-\frac{1}{2}} D_n(\tau_n)$ tends to zero for τ_n approximately determined by ${\tau_n}^2 = CD(\tau_n)$ as in Theorem 1.

It is also possible, in this realm of ideas, to prove that every pairwise imperfect infinitely divisible experiment is a direct product of a gaussian and a Poisson experiment; see Le Cam [1986, p. 168] and Janssen, Milbrot and Strasser [1985].

4.4 Historical Remarks

The so-called "gaussian" or "normal" distribution occurs in the work of de Moivre [1733]. This seemed to have been in a probabilistic situation not directly involving problems of estimation or testing. The gaussian experiments occur as "experiments" in a paper of C. F. Gauss [1809]. Gauss gave a "justification" of the method of least squares of Legendre [1805] in which he observes that the family obtained by shifting the de Moivre density (the normal density) is the only shift family for which the average of the observations is what we now call the maximum likelihood estimate. This last bit of terminology is due to Fisher. Gauss himself used the mode of the posterior density for a "flat" prior, namely, the Lebesgue measure on $(-\infty, +\infty)$.

Gaussian experiments also occur around that same year in the works of Laplace. He had proved a central limit theorem for sums of small independent variables and was very interested in "Le milieu qu'il faut prendre entre plusieurs observations." Laplace also proved a form of what is now called the Bernstein-von Mises theorem (see Chapter 8), according to which posterior distributions tend to be approximable by gaussian distributions.

It appears possible that the *shift* character of the experiments under consideration was an incidental feature, probably inadvertently introduced by the notation of the astronomers of the time with their "equations de condition." It was severely criticized by J. Bertrand and H. Poincaré. Both of these authors insisted that one must consider densities $f(x, \theta)$ and not restrict oneself to the shift form $f(x-\theta)$. Following Gauss's argument, they came within a hair's breadth of recommending what is now called "exponential families."

Because of their mathematical tractability, gaussian experiments soon became the main stay of statistics. Even when the covariance matrices had to be estimated, leading to multivariate analysis, they were studied in great detail. It is only in the early 1950s that one realized that although "mathematically tractable," they were not statistically so simple.

The work of Charles Stein [1956] showed that for the usual loss functions and for a gaussian shift experiment indexed by an entire euclidean space of dimension three or more, the mean of the observations was not admissible. James and Stein [1961] go further. After the work of L.D. Brown [1966], one could hardly doubt the general prevalence of inadmissibility for fairly general loss functions and so-called "equivariant" estimates. The phenomenon is not particular to gaussian shift experiments, but it is spectacularly counterintuitive there.

It has been said, but not often enough, that there are *no* known solid arguments compelling the statistician to use equivariant estimates. The arguments by analogy with changes of physical units are specious. If a European gives you estimates of fuel consumption in liters per 100 kilometers, the estimate can readily be converted to miles per gallon without implying anything at all about the mathematical formula or the procedure that was used for the initial estimate. Still, in Chapter 6, we shall encounter a form of asymptotic equivariance implied by the passage to the limit, almost everywhere in finite-dimensional situations. The implication should probably be studied more carefully.

Gaussian experiments have led to other recent and important studies. One of them, also by Charles Stein [1956] was of a "semiparametric" nature, pointing out that if one wants to estimate a one-dimensional function in a gaussian experiment indexed by a full Hilbert space, there is a "most difficult" one-dimensional problem whose solution gives that of the original problem. This was later extended by D. Donoho to experiments indexed by certain convex symmetric subsets of Hilbert space, see e.g. Donoho [1997]. This is more difficult.

Estimates that are better than the equivariant ones have been proposed by many authors after the original construction of C. Stein. They include the ridge estimates in regression analysis and the threshold estimates of Donoho for wavelet expansions.

As already mentioned in the text, Poisson experiments occur in a variety of fields. There are two recent books, one by D. Aldous [1989] and another by A. Barbour, L. Holst and S. Janson [1992], that describe a large variety of probabilistic considerations involving them. They are also involved in many constructions of random measures or so-called "point processes"; see Bochner [1955] and Kallenberg [1983], among others. However, there does not seem to exist a body of theory about the properties of Poisson experiments as statistical structures. There do exist some statistical statements, derived from their infinite divisibility, as, for instance, in Le Cam [1986, p. 482 *sqq*]. However, their study has been slight compared to the study of gaussian experiments, in spite of the fact that one can view the gaussian case as a "degenerate" limit of the Poisson situation.

5

Limit Laws for Likelihood Ratios

5.1 Introduction

In this chapter we shall consider a double sequence $\{\mathcal{E}_{n,j}; j = 1, 2, ...; n = 1, 2, ...\}$ of experiments $\mathcal{E}_{n,j} = \{p_{t,n,j}; t \in \Theta\}$. Let $\mathcal{E}_n$ be the direct product in j of the $\mathcal{E}_{n,j}$. That is, $\mathcal{E}_n$ consists of performing the $\mathcal{E}_{n,j}, j = 1, 2...$ independently of each other. The measures that constitute $\mathcal{E}_n$ are the product measures $P_{t,n} = \prod_j p_{t,n,j}$. It will usually be assumed that j runs through a finite set.

We propose studying the limiting behavior of loglikelihood ratios of the type $\Lambda_n = \log(dP_{t,n}/dP_{s,n})$ when, as n tends to infinity, the "information" given by each individual experiment becomes small. A result of this type already occurred in Chapter 4 in the discussion of "Poissonization." It is possible, but delicate to deduce most of the results that on the behavior of Λ_n from the remarks of Chapter 4, but more direct statements are practically useful. As before, $\Lambda_n = \log(dP_{t,n}/dP_{s,n})$ will always be defined as the Radon-Nikodym derivative of the part of $P_{t,n}$ that is dominated by $P_{s,n}$. Since we are concerned with distributions under $P_{s,n}$, Λ_n is arbitrarily defined on the singular part of $P_{t,n}$ with respect to $P_{s,n}$.

5.2 Auxiliary Results

In this section, we have collected a number of results and descriptions of methods that will be useful in subsequent developments of the present chapter. We are concerned with finite sums of independent random variables $\{X_j; j = 1, 2, ...\}$ unless stated otherwise. We give a proof of the central limit theorem from scratch, emphasizing approximations more than limits. The technique is not new. For the gaussian case, it goes back to Lindeberg [1920]. The rest is imitated from the writings of Paul Lévy of the mid-thirties. Except in Section 3, it does not use special machinery, such as characteristic functions. The proofs were intended to introduce the Poisson approximation as soon as possible and to make it easy to obtain statements free of the complications created by expectations of *truncated* variables.

In the special case of gaussian limits, we do not rely on the famous Lindeberg-Feller condition. This condition is necessary and sufficient for convergence to a gaussian limit *whose variance is the limit of the variances of the sums* $\sum_j X_{n,j}$. This is a special feature that will be shown to be necessary for contiguity if one sums the centered *square roots* of densities. If one sums the densities themselves or their logarithms, the situation is very different. The procedure of proof used makes it not only possible but easy to pass from one version of such sums to another.

5.2.1 Lindeberg's Procedure

This is the procedure used by Lindeberg to give an elementary proof of the Central Limit Theorem in the gaussian case. It uses nothing more advanced than Taylor's formula. It is based on the remark that given a function f evaluated at $(x_1, x_2, \ldots, x_n)$ one can pass to its evaluation at $(y_1, y_2, \ldots, y_n)$ by changing one variable at a time. Here the technique will be applied to functions of a sum $\sum_j X_j$ of n independent random variables. The X_j will be replaced by other independent variables Y_j, independent of the X_j.

To proceed let $R_k = \sum_{j<k} X_j + \sum_{j>k} Y_j$, empty sums being equal to 0. Then

$$\begin{aligned} f(\sum_j X_j) &- f(\sum_j Y_j) \\ = \sum_k [f(R_k + X_k) &- f(R_k + Y_k)]. \end{aligned} \tag{5.1}$$

Each term in the last sum is a difference of values of f at places that differ only by the substitution of one Y_k for one X_k. This gives the following.

Proposition 1. *Assume that all the X_j and Y_j are independent. Assume also that $EX_j = EY_j = 0$ and $EX_j^2 = EY_j^2 = \sigma_j^2$. Let $\beta_j = E|X_j|^3 + E|Y_j|^3$. Then for any real-valued f whose second derivative f'' satisfies the Lipschitz condition $|f''(x) - f''(y)| \leq A|x - y|$ one has*

$$|Ef(\sum_j X_j) - Ef(\sum_j Y_j)| \leq \frac{A}{2} \sum_j \beta_j. \tag{5.2}$$

Proof. Consider one of the terms $f(R_k + X_k) - f(R_k + Y_k)$ in (5.1). Expanding such a term around R_k one obtains

$$\begin{aligned} f(R_k + X_k) &= f(R_k) + X_k f'(R_k) + (1/2)X_k^2 f''(R_k) + W_{k,1}, \\ f(R_k + Y_k) &= f(R_k) + Y_k f'(R_k) + (1/2)Y_k^2 f''(R_k) + W_{k,2}, \end{aligned}$$

where the W are remainder terms of the expansions. For example, $W_{k,1} = (1/2)[f''(\xi) - f''(R_k)]X_k^2$ for some ξ between R_k and $R_k + X_k$.

Taking a difference and expectations, one sees that, since the X_k and Y_k are independent of R_k, all the terms disappear except the remainder terms. For these note that

$$|W_{k,1}| \leq \frac{1}{2} A X_k^2 |\xi - R_k| \leq \frac{A}{2} |X_k|^3;$$

similarly for $W_{k,2}$.

Taking expectations and summing over k gives the desired result. □

At first sight this theorem does not seem to have much content, but it does. To extract some, let us introduce the Kolmogorov norm

$$\|P - Q\|_K = \sup_x |P(-\infty, x] - Q(-\infty, x]|,$$

where P and Q are probability measures.

Also, to simplify the notation let $S = \sum_{j=1}^{n} X_j$ and $T = \sum_{j=1}^{n} Y_j$.

Proposition 2 (Lindeberg). *Assume that the X_j for $j = 1, \ldots, n$, are independent with $EX_j = 0, EX_j^2 = \sigma_j^2$. Assume also that for each j one has $|X_j| \leq a$. Let $s^2 = \sum_j \sigma_j^2$, let P be the law of S, and let G be the normal law with mean 0 and standard deviation s. Then*

$$\|P - G\|_K \leq C(\frac{a}{s})^{\frac{1}{4}} \tag{5.3}$$

where C is a universal constant.

Proof. We shall apply our previous proposition to a function f obtained from a smooth function g, say, a beta distribution function. Let $g'(x)$ be the beta density $B(m,m)^{-1}x^{m-1}(1-x)^{m-1}$ for $0 \leq x \leq 1$ and zero otherwise. Then $g(x) = \int_{-\infty}^{x} g'(t)dt$. One has $g(x) = 0$ for $x < 0$ and $g(x) = 1$ for $x > 1$. Furthermore, if $m \geq 6$, the first four derivatives are continuous and bounded. By a change of scale, we take $f(x) = g(\tau x)$. Then $f(x) = 0$ for $x \leq 0$ and $f(x) = 1$ for $x \geq 1/\tau$. Letting $S = \sum X_j, T = \sum Y_j$, as indicated, we can write

$$\begin{aligned} Pr[S - x > 0] &\leq Ef(S - x + \frac{1}{\tau}) \\ &\leq Ef(T - x + \frac{1}{\tau}) + C_g \tau^3 a s^2 \\ &\leq P[T - x > 0] + P[\frac{-1}{\tau} \leq T - x < 0] \\ &\quad + C_g \tau^3 a s^2. \end{aligned}$$

For the gaussian T one has $P[0 \leq T - x < \frac{1}{\tau}] < 1/s\tau\sqrt{2\pi}$. The inequality follows by selecting an optimal value for τ and applying the same reasoning to the inequality in the opposite direction.

Note that C_g depends only on the choice of the smooth function g, so that the constant is indeed "universal." It is not worth computing, since the term $(a/s)^{\frac{1}{4}}$ is far from the best possible. An argument of Esseen, using Fourier transforms, would readily give a bound of the type $C(a/s)$. This does not mean that

Lindeberg's argument should be forgotten; it is simple. What is more, it applies to many cases where other procedures do not work easily. For instance, it applies without modifications to Hilbert space valued variables. Paul Lévy [1937] used it for martingales. Even for the case of independent variables, we shall need a modification as follows.

Consider independent variables X_j as above. Let F_j be the distribution of X_j. Replace F_j by its poissonized version $\exp[F_j - I]$, the distribution of $Y_j = \sum_i^{\nu_j} X_{j,i}$, where the ν_j are independent Poisson variables with $E\nu_j = 1$ and $X_{j,i}$ are independent with a common distribution F_j. Here I, the Dirac measure that has mass one at 0, denotes the identity in the convolution algebra. (For further discussion, see Section 5.2.5.)

As before we shall let $S = \sum_{j=1}^n X_j$ and $T_p = \sum_{j=1}^n Y_j$.

Proposition 3 (Poissonization). *Let X_j for $j = 1, \ldots, n$, satisfy the hypothesis of Proposition 2. Let f be a function whose fourth derivative is bounded by a number B. Then*

$$|Ef(S) - Ef(T_p)| \leq Ba^2 s^2.$$

Proof. The argument follows almost exactly that of our previous proposition. Here the X_j and Y_j have the same first third moments. The third absolute moment of Y_j is not readily computable. So we have used a function f having four derivatives. Making a Taylor expansion to three terms with a remainder in the fourth derivative, one sees that the first three terms of the expansion cancel. The remainder term can be bounded using fourth moments $EX_j^4 \leq a^2\sigma_j^2$ and $EY_j^4 = EX_j^4 + 3\sigma_j^4$.

This inequality can be converted into an inequality for the Lévy distance between the laws of S and T_p. It can also yield an inequality for the Kolmogorov distance, using bounds for the concentration function $\sup_x Pr[0 \leq T_p - x \leq 1/\tau]$ as was done for Proposition 2 (Lindeberg's theorem). However, we have not derived acceptable bounds for such concentration functions. More discussion, and some bounds, can be found in Le Cam [1986, chapter 15].

5.2.2 Lévy Splittings

Lévy introduced a method that, together with the Lindeberg procedure, allows a simple proof of central limit theorems for the case of sums of independent variables. It "splits" the large values of the variables from the small ones. Then the limiting distributions are investigated separately for the two parts. The splitting is done as follows.

If F is a probability measure on the line, one can always write it as a sum of two positive nonnull measures, say, $F = F^1 + F^2$. Renormalizing each of the F^i, one obtains two new probability measures, say, M and N. This leads to writing F as a convex combination $F = (1-\alpha)M + \alpha N$. To this corresponds a construction in terms of independent random variables: Let ξ be a random variable such that

$$P[\xi = 1] = \alpha = 1 - P[\xi = 0].$$

Let U' have distribution M and V' have distribution N. Suppose that all three variables $\xi, U' and V'$ are independent, and consider the sum $X = (1-\xi)U' + \xi V'$. Simple algebra shows that the distribution of X is precisely F.

If one considers a sequence $\{X_j; j = 1, 2, \ldots\}$ of independent variables, with distribution F_j for X_j. This splitting can be applied to each X_j, giving a representation

$$X_j = (1-\xi_j)U'_j + \xi_j V'_j,$$

in distribution where all the variables are mutually independent.

This is often done by taking a number a and

$$F_j = (1-\alpha_j)M_j + \alpha_j N_j,$$

where $\alpha_j = Pr[|X_j| > a]$ and M_j and N_j are, respectively, the conditional distributions of X_j given $|X_j| \leq a$ and given $|X_j| > a$. In such a case the expectation $m_j = EU'_j$ exists. More generally, for an arbitrary splitting, F^1 and F^2, we assume the existence of m_j and introduce new variables:

$$U_j = U'_j - m_j \quad \text{and} \quad V_j = V'_j - m_j.$$

Then

$$X_j = m_j + (1 - \xi_j)U_j + \xi_j V_j$$

in distribution.

Let us go a step further, introducing another set of random variables $\{\eta_j\}$ independent among themselves and independent of our previous variables, but such that ξ_j and η_j have the same distribution. Then, as far as the distributions are concerned, one can write

$$\sum_j [X_j - m_j] = \sum_j (1 - \eta_j)U_j + \sum_j \xi_j V_j + \sum_j (\eta_j - \xi_j)U_j. \quad (5.4)$$

The point of the operation is simple: *The first two sums* $\sum(1 - \eta_j)U_j$ *and* $\sum \xi_j V_j$ *are independent. Furthermore, the cross term* $\sum(\eta_j - \xi_j)U_j$ *is "negligible" compared to* $\sum(1 - \eta_j)U_j$ *whenever* α *is "small".* This can be guessed by noting that the variance of the cross term never exceeds 2α times the variance of $\sum(1 - \eta_j)U_j$. A better description is given in Le Cam [1986, p. 412, 436]. We shall not give the argument in full here. It will be needed only in a special case of Proposition 6 later. Here, as in Proposition 6, we have split the variables X_j according to the values they take; one can use other criteria. The main idea is to write the distribution F_j of X_j in the form $F_j = (1 - \alpha_j)M_j + \alpha_j N_j$ where the α_j are small. For example, it will happen that X_j is a function $X_j = f_j(Y_j)$ of some other variable Y_j. One can split the X_j according to the values taken by Y_j, letting, for instance, N_j be the conditional distribution of X_j given that $|Y_j| > b_j$. This flexibility will be used in Section 5.3.

5.2.3 Paul Lévy's Symmetrization Inequalities

Proposition 4. *Let* $X_j; j = 1, 2, ..., m$, *be independent random variables. Let* $S_k = \sum_{j \le k} X_j$. *Let* $T_k = \sum_{k<j} X_j$, *for* $k = 1, \ldots, m - 1$, *have medians zero. Then, for* $a \ge 0$,

$$Pr[\sup_k |S_k| \ge a] \le 2Pr[|S_m| \ge a].$$

Proof. Let k be the first index such that $S_k \geq a$. The remaining sum T_k has median zero. Thus $S_m = S_k + T_k$ has probability at least 1/2 of being as large (or as small) as S_k. The argument also applies to nonindependent X_j as long as, *given the past*, $S_1, \ldots, S_k$, the T_k have conditional zero medians. □

Note 1. This is known as the symmetrization inequality although only the condition of zero medians is needed in the proof. Zero medians are satisfied by the symmetrized random variables $X_j - X'_j$, where X'_j is an i.i.d. copy of X_j. They will be used later.

5.2.4 Conditions for Shift-Compactness

In the sequel, we shall consider "compactness" or "convergence" properties of probability measures P_n, P. This assumes the choice of a topology. Here we shall use the "ordinary topology," where $P_n \longrightarrow P$ if $\int f dP_n \longrightarrow \int f dP$ for every bounded continuous function f. The topology can be induced by the Lévy metric or the dual Lipschitz norm defined in Section 2.2. See Dudley [1989, chapter 11] for more details. Note that the ordinary topology is considerably weaker than the topology induced by the Kolmogorov norm used in Proposition 2.

Consider a sequence S_n of sums $S_n = \sum_j X_{n,j}$ where for every fixed n, $X_{n,j}$ are independent random variables. If there are constants a_n such that the distributions of $S_n - a_n$ form a relatively compact sequence on the line, the S_n are usually called "shift-compact," although such an appellation should really be attached to their laws. Note that relative compactness was defined in Section 3.1. It is a necessary condition for weak convergence of the distributions of S_n.

If the S_n are shift-compact one can take for centerings a_n the medians of S_n. Then a recourse to Paul Lévy's symmetrization inequalities in Proposition 4 shows that shift-compactness of S_n is equivalent to relative compactness of sums $S_n - S'_n$ where S'_n is independent of S_n and has the same distribution as S_n. Thus, we shall only give conditions for relative compactness applicable to symmetric variables $X_{n,j}$.

Proposition 5. *Let the $X_{n,j}$ be symmetrically distributed about zero and independent. In order for the distributions of $S_n = \sum_j X_{n,j}$ to form a relatively compact sequence it is necessary and sufficient that the following two conditions be satisfied:*

(i) For $a > 0$ let $X_{n,j,a} = X_{n,j}I[|X_{n,j}| < a]$. Then, for each a finite, the sum of the variances of the $X_{n,j,a}$ is bounded by some $b(a) < \infty$.

(ii) For every $\epsilon > 0$ there is a finite a such that

$$\sum_j Pr[|X_{n,j}| \geq a] \leq \epsilon.$$

Proof. Let us first prove the necessity of (ii). For this, let $S_{n,k} = \sum_{j=1}^{k} X_{n,j}$. The relative compactness implies, for a given ϵ, the existence of a number a such that $Pr[|S_n| \geq a/2] < \epsilon/2$. By Paul Lévy's inequalities, this implies

$$Pr[\sup_k |S_{n,k}| \geq \frac{a}{2}] < \epsilon.$$

Therefore, taking differences, $X_{n,k} = S_{n,k} - S_{n,k-1}$, we obtain

$$Pr[\sup_j |X_{n,j}| \geq a] < \epsilon.$$

The result follows (modifying ϵ a bit) from the inequality

$$1 - Pr[\sup |X_{n,j}| \geq a] = \prod (1 - \alpha_{n,j}) \leq \exp(-\sum \alpha_{n,j}),$$

where $\alpha_{n,j} = P[|X_{n,j}| \leq a]$.

To prove that relative compactness implies (i), apply the preceding to a small ϵ, say, $\epsilon \leq 10^{-2}$. This provides a number a. Take $b > a$ and suppose that σ_n^2, the variance of $\sum_j X_{n,j,b}$, is unbounded. Taking a subsequence if necessary, one can assume that σ_n tends to infinity. Then, by Proposition 2, the law of $\sum_j X_{n,j,b}$ becomes arbitrarily close in Kolmogorov norm to a gaussian distribution of standard deviation σ_n. This gaussian sequence is definitely not relatively compact, and the lack of compactness cannot be remedied by adding a term that has probability at least $1 - 10^{-2}$ of being zero.

Conversely, the conditions (i) and (ii) imply the relative compactness of S_n. To see this, let us note that for any positive ϵ and a,

$$\begin{aligned} P[|S_n| \geq 2\epsilon] &\leq P[|\sum_j X_{n,j,a}| \geq \epsilon] \\ &+ P[|\sum_j X_{n,j} I[|X_{n,j}| > a]| \geq \epsilon]. \end{aligned}$$

Consider the sum $\sum_j X_{n,j} I[|X_{n,j}| > a]$. It differs from zero only when at least one of the indicators $I[|X_{n,j}| > a]$ takes the value one. By condition (ii) of the proposition, one can choose the number a in such a way that

$$Pr[\sup_j I[|X_{n,j}| > a] > 0] \leq \sum_j Pr[|X_{n,j}| > a] \leq \epsilon.$$

According to condition (i), the variance of the sum $\sum_j X_{n,j,a}$ is bounded by some $b < \infty$. Thus, by Chebyshev, if t is so chosen that $b^2/t^2 < \epsilon$ then $Pr[|\sum_j X_{n,j}| \geq \max(t, a)|] < 2\epsilon$. This gives the desired "tightness" of S_n, hence relative compactness. □

5.2.5 A Central Limit Theorem for Infinitesimal Arrays

Consider random variables $X_{n,j}, j = 1, 2, \ldots$ as in Section 2.1, independent for each given n, but under the condition that they are "infinitesimal" or, in Loève's terminology, "uniformly asymptotically negligible"(UAN), that is, for every $\epsilon > 0$ the quantity $\sup_j Pr[|X_{n,j}| > \epsilon]$ tends to zero as n tends to infinity.

Take such an infinitesimal array and a $\tau > 0$. Let $Y_{n,j} = X_{n,j} - m_{n,j}$ where $m_{n,j}$ is the conditional expectation of $X_{n,j}$ given that $|X_{n,j}| < \tau$. Let $G_{n,j}$ be the distribution of $Y_{n,j}$. Let $L_n = \sum_j G_{n,j}$, except that the mass at zero has been removed.

We shall, in Proposition 6, prove a central limit theorem for the double array $\{Y_{n,j}\}$. Before stating it, let us elaborate on

Poisson sums and poissonization. If $G_{n,j}$ is the distribution of $Y_{n,j}$, its poissonized version is the distribution of a sum $\sum_{i=1}^{N} \xi_i$, where the ξ_i are i.i.d. with distribution $G_{n,j}$ and N is a Poisson variable independent of the ξ_j with expectation $EN = 1$. It is convenient to express this poissonization operation in the convolution algebra of measures. The poissonized version of $G_{n,j}$ is then the exponential $\exp(G_{n,j} - I)$, where I is the probability measure with mass 1 at 0. The convolution exponential of a finite signed measure μ is defined by the usual series expansion

$$\exp \mu = e^{\mu} = \sum_{k=0}^{\infty} \frac{\mu^k}{k!},$$

where μ^k is the kth convolution of μ and $\mu^0 = I$. Note that I is the identity of the convolution algebra and that, for a number a, convoluting by $\exp aI$ is the same thing as multiplying by e^a. Thus $\exp(G_{n,j} - I) = e^{-1} \exp(G_{n,j})$. Note that if $G_{n,j}$ has a mass, say, γ at 0, one can let $G'_{n,j} = G_{n,j} - \gamma\delta_0$. Then $G_{n,j} - I = G'_{n,j} - \|G'_{n,j}\|$. Thus the mass at zero plays no role and can be assumed to have been removed, as was done for L_n.

With these definitions, the infinitely divisible distribution accompanying the array $\{Y_{n,j}\}$ is $\exp\{\sum_j [G_{n,j} - I]\}$ equal to $\exp[L_n - \|L_n\|]$. The measure L_n is called the Lévy measure of the distribution $\exp[L_n - \|L_n\|]$.

Recall that we have let $Y_{n,j} = X_{n,j} - m_{n,j}$ where $m_{n,j}$ is the conditional expectation of $X_{n,j}$ given $|X_{n,j}| < \tau$.

Proposition 6. *Assume that the array $\{X_{n,j}\}$ is infinitesimal and the sum $\sum_j X_{n,j}$ shift-compact. Then, as n tends to infinity, the Lévy or dual Lipschitz distance between the distribution of $\sum_j Y_{n,j}$ and the infinitely divisible distribution* $\exp[L_n - \|L_n\|]$ *tends to zero, as n tends to infinity.*

Proof. Write $X_{n,j} = (1 - \xi_{n,j})U'_{n,j} + \xi_{n,j}V'_{n,j}$ where $U'_{n,j}$ has a distribution equal to the conditional distribution of $X_{n,j}$ given $|X_{n,j}| < \tau$ and $\xi_{n,j}$ equals in distribution to $I[|X_{n,j}| \geq \tau]$ as was done for the Lévy splitting. This gives as in (5.4)

$$Y_{n,j} = X_{n,j} - m_{n,j}$$

$$\begin{aligned} &= (1-\xi_{n,j})U_{n,j} + \xi_{n,j}V_{n,j} \\ &= (1-\eta_{n,j})U_{n,j} + \xi_{n,j}V_{n,j} + (\eta_{n,j} - \xi_{n,j})U_{n,j} \end{aligned}$$

where now $U_{n,j} = U'_{n,j} - m_{n,j}$ has expectation zero. By shift-compactness, the sum of the variances of the $U_{n,j}$ stays bounded (see Proposition 5), and by the UAN condition, $\sup_j P[\xi_{n,j} = 1] \to 0$. Therefore the cross term $\sum_j (\eta_{n,j} - \xi_{n,j})U_{n,j}$ tends to zero in quadratic mean and can be neglected. One is reduced to consider the sum of two independent terms $\sum_j (1-\eta_{n,j})U_{n,j}$ and $\sum_j \xi_{n,j}V_{n.j}$. We shall show separately that each such term has a distribution approximable by its accompanying infinitely divisible distribution. The first sum is subject to the application of Lindeberg's procedure as was explained in Proposition 3. For the second sum, let

$$\beta_{n,j} = Pr[\xi_{n,j}V_{n.j} \neq 0].$$

By the shift-compactness and Proposition 5, $\sum_j \beta_{n,j}$ stays bounded. Hence according to the UAN condition, $\sum_j \beta^2_{n,j}$ tends to zero as n tends to infinity. A simple coupling argument, replacing Bernoulli variables by Poisson ones, yields the infinite divisible approximation for the second sum.

Explicitly, let $Z_{n,j,k}$ for $k = 1, 2, \ldots,$ be i.i.d. random variables whose distribution is the conditional distribution of $\xi_{n,j}V_{n,j}$ given that $\xi_{n,j}V_{n,j} \neq 0$. Thus $\xi_{n,j}V_{n,j}$ can be represented in distribution as $\omega_{n,j}Z_{n,j,1}$ where $\omega_{n,j}$ is a Bernoulli variable independent of $Z_{n,j,1}$ and such that $Pr[\omega_{n,j} = 1] = \beta_{n,j}$. Consider also a Poisson sum $\sum_{k=0}^{N_j} Z_{n,j,k}$ where N_j is an independent Poisson variable with $EN_j = \beta_{n,j}$ and $Z_{n,j,0} \equiv 0$. Note that the Poisson sum and $\omega_{n,j}Z_{n,j,1}$ share the same $Z_{n,j,1}$ and are dependent. For coupling, we construct a joint distribution of $\omega_{n,j}$ and N_j such that their marginal distributions are equal to the preceding specification. This can be achieved by assigning

$$\begin{aligned} P[(\omega_{n,j}, N_j) = (0,0)] &= 1 - \beta_{n,j}, \\ P[(\omega_{n,j}, N_j) = (1,0)] &= e^{-\beta_{n,j}} - (1-\beta_{n,j}), \\ P[(\omega_{n,j}, N_j) = (1,n)] &= \frac{e^{-\beta_{n,j}}\beta^n_{n,j}}{n!}, \quad \text{for } n \geq 1. \end{aligned}$$

Consequently,

$$P[\sum_{k=0}^{N_j} Z_{n,j,k} \neq \omega_{n,j} Z_{n,j,1}] = \beta_{n,j}(1 - e^{-\beta_{n,j}}) \leq \beta_{n,j}{}^2.$$

It follows that the L_1-distance between the distributions of $\sum_j \xi_{n,j} V_{n,j}$ and that of the Poisson sum $\sum_j \sum_k^{N_j} Z_{n,j,k}$ tends to zero as $n \to \infty$. Hence the desired result follows by convolution of the distributions of the two sums, and by noting that, for the poissonized versions, convolution corresponds to addition of the Lévy measures. □

Remark 1. The preceding proof makes essential use of a shift-compactness assumption. Following the work of Kolmogorov [1956] and Le Cam [1965] showed that, at least for the Lévy distance, the statement of Proposition 6 is valid generally, without the shift-compactness assumption. For details, see the monograph by Arak and Zaitsev [1988].

5.2.6 The Special Case of Gaussian Limits

In the preceding section we introduced infinitesimal arrays with approximation of sums by infinitely divisible distributions. The case where these infinitely divisible distributions are gaussian exhibits special features that will be used repeatedly in the sequel. The main feature can be stated as follows. In an infinitesimal array $X_{n,j}$, for every $\tau > 0$ one has $\sup_j Pr[|X_{n,j}| > \tau]$ tending to zero as $n \to \infty$. In the case of gaussian limits, the sup and the Pr are interchanged, that is, $Pr[\sup_j |X_{n,j}| > \tau]$ tends to zero, and this is equivalent to $\sum_j P[|X_{n,j}| > \tau]$ tending to zero. Explicitly

Proposition 7. *Assume that the $X_{n,j}$ form an infinitesimal array and that the $S_n = \sum_j X_{n,j}$ are shift-compact. Let F_n be the distribution of S_n. There are gaussian distributions G_n such that the Lévy distance between F_n and G_n tends to zero if and only if for every $\tau > 0$ the sums $\sum_j Pr[|X_{n,j}| > \tau]$ tend to zero as $n \to \infty$.*

Proof. According to Lindeberg's theorem (Proposition 2), it is enough to prove the necessity of the condition.

If the F_n admit gaussian approximations, so do the distributions of symmetrized sums $\sum_j (X_{n,j} - X'_{n,j})$. Thus it is sufficient to consider the case of variables $X_{n,j}$ that have symmetric distributions about zero. Also, the case where the G_n have variances tending to zero follows immediately from Proposition 4. Thus, taking subsequences as needed, we shall assume that the F_n tend to a gaussian limit G. By rescaling, we take G to be the distribution of a standard gaussian variable W with $EW = 0$, $EW^2 = 1$, and $EW^4 = 3$.

Let L_n be the sum of the distributions of the $X_{n,j}$ (with masses at zero deleted, if one so wishes). By assumption, the sequence S_n is relatively compact, it follows from (ii) of Proposition 5 that for each $\epsilon > 0$ there is an a depending on ϵ such that the mass of L_n situated outside $(-a, a)$ never exceeds ϵ. Thus, one can assume without changing limits of L_n that L_n is supported by a finite interval $[-a_n, a_n]$ where a_n tends to infinity arbitrarily slowly. Now we apply Proposition 6. Note that $m_{n,j} = 0$ since $X_{n,j}$ are symmetric. Then by Proposition 6, F_n is approximated by the convolution exponential $H_n = \exp[L_n - \|L_n\|]$ whose Laplace transforms exist. They are related by the equation

$$\int e^{tx} H_n(dx) = \exp\{\int [e^{tx} - 1] L_n(dx)\}, \quad -\infty < t < \infty, \tag{5.5}$$

as can be seen from the Poisson formulation. Because by assumption $X_{n,j}$ and hence L_n is symmetric around zero, the e^{tx} can be replaced by $\cosh(tx)$. Then the integral in the exponential on the right becomes $\int [\cosh(tx) - 1] L_n(dx)$, an integral of a positive function.

Let us assume first that for some positive t these Laplace transforms stay bounded in n. Then all the moments of H_n converge to those of G, and so do the cumulants. In particular, the fourth cumulant of H_n tends to the fourth cumulant of G that is 0. However the *cumulants* of H_n are the *moments* of L_n. Thus, $\int x^4 L_n(dx)$ tends to zero, implying the desired conclusion.

To get the general result without assuming the existence of a positive t, take a value $a : 0 < a < \infty$. Split the measure L_n as

$L_{n,1}+L_{n,2}$ where $L_{n,1}$ is the part of L_n carried by $[-a, a]$ and $L_{n,2}$ the part outside of $[-a, a]$. This gives H_n as a convolution $H_n = H_{n,1} \star H_{n,2}$. One can assume that $H_{n,1}$ tends to a limit, say, K and that $L_{n,2}$ tends to a limit M. Then $H_{n,2}$ tends to the Poisson exponential $\exp[M - \|M\|]$. In this case $\int[\cosh(tx) - 1]L_{n,1}(dx)$ stays bounded, because for $|tx| \leq 1$ there is a C such that $0 \leq \cosh(tx) - 1 \leq Ct^2x^2$. Therefore the cumulants of $H_{n,1}$ tend to those of K. The fourth cumulant of K is therefore positive or zero. The fourth cumulant of $\exp[M - \|M\|]$ is also positive or zero. Their sum, the cumulant of G, is zero, so both must vanish. This implies that M must vanish, completing the proof of the proposition. □

Note 2. As can be seen, the proof uses only the fact that the fourth cumulant of G is zero. It is an elaboration on the fact that if an infinitely divisible distribution has four moments, its fourth cumulant is nonnegative, vanishing only if the distribution is gaussian.

Note 3. In the preceding we mentioned convergence of distributions of sums to gaussian, but we have not mentioned, except in Proposition 2, the relation between the variance of the sums and the variance of the limiting distribution. In particular, we have not mentioned the classical Lindeberg condition. It assumes the existence of means and variances, and therefore it does not apply directly to the $X_{n,j}$ of Proposition 7. If applicable to the $X_{n,j}$, it would read as follows. Let $m_{n,j} = E\{X_{n,j} | |X_{n,j}| \leq \tau\}$, then

$$\sum_j E(X_{n,j} - m_{n,j})^2 I[|X_{n,j} - m_{n,j}| > \tau] \to 0.$$

Note that we have not normalized it by the variance of the sum of the $X_{n,j}$, since it may be infinite in Proposition 7. The classical Lindeberg condition is necessary and sufficient for the occurrence of gaussian limits *whose variance is the limit of the variances of the sums* $\sum_j X_{n,j}$. Our Proposition 7 says nothing about the limit of variances. This is on purpose. Typically, the variance of the limiting distribution may be strictly smaller than the limit of variances of sums. A case of equality will be encountered in Section 3 in connection with contiguity.

5.2.7 Peano Differentiable Functions

Sometime before 1900, mathematicians felt the need to make explicit the role of derivatives in approximations of functions near a point. This resulted in definitions from which we shall retain the definitions of Peano and Fréchet.

For Peano derivatives, consider a function f from some subset S of the real line taking values in $[-\infty, +\infty]$. Assume that $0 \in S$. Then f is Peano differentiable of order one at 0 if

$$f(x) - f(0) = ax + xR_1(x)$$

where $R_1(x)$ tends to zero as x tends to zero.

The function f is differentiable of order two at 0 if

$$f(x) - f(0) = ax + \frac{1}{2}bx^2 + x^2 R_2(x)$$

where $R_2(x)$ tends to zero as x tends to zero.

(To be precise, we should specify that the $R_i(x)$ tend to zero as x tends to zero *while staying in* S. In other words, the approximations to f are only where f is defined.) Peano differentiability of order two does not require the existence of the *first* derivative in a neighborhood of zero and therefore may exist more generally than is the case for a second derivative obtained by taking a derivative of the first derivative. This is immaterial for our purpose. We only want to emphasize the approximation aspect.

The Fréchet derivative is defined similarly, but for a function f from a normed space $\mathcal{X}$ to a topological vector space, $\mathcal{Y}$. The function f has Fréchet derivative A at 0 if

$$f(x) - f(0) = Ax + \|x\| R(x)$$

where A is a continuous linear operation from $\mathcal{X}$ to $\mathcal{Y}$ and where $R(x)$ tends to zero as $\|x\|$ does. The linear operation A is usually assumed to be a continuous operation but there have been discussions on that point. We shall use Fréchet derivatives, for example, under the name of "derivatives in quadratic mean" in Chapter 7.

In studying limit laws for likelihood ratios we shall use functions f that are Peano differentiable of order two at 0 with the added restriction that $f(0) = 0$. We shall call them functions of the class $P[0, 2]$. Their role is illustrated by the following simple observation.

Proposition 8. *Let $\{X_{n,j}\}$ be a double array of random variables such that as $n \to \infty$, the sum $\sum_j \|X_{n,j}\|^2$ stays bounded in probability and $\sup_j \|X_{n,j}\|$ tends to zero in probability. Let f be of class $P[0, 2]$, with first and second derivatives a and b. Then*

$$\sum_j f(X_{n,j}) - \sum_j [aX_{n,j} + \frac{1}{2}b(X_{n,j})^2]$$

tends to zero in probability.

Proof. The proof is immediate. The difference in the statement equals $\sum_j \|X_{n,j}\|^2 R_2(X_{n,j})$. It does not exceed

$$\sum_j \|X_{n,j}\|^2 \sup_j \|R(X_{n,j}\|,$$

which tends to zero in probability. □

Note that the result does not assume any form of independence of the $X_{n,j}$, although we shall mostly use it in that case. It is particularly interesting when the sum of the square terms tends to a nonrandom limit, so that the random part of $\sum_j f(X_{n,j})$ boils down to the sum $a \sum_j X_{n,j}$.

The conditions of Proposition 8 can be restated as follows:

C_1. $\sup_j |X_{n,j}|$ tends to zero in probability.

C_2. The sums $\sum_j X_{n,j}^2$ stay bounded in probability.

The reader will note the connection between C_1 and the result of Proposition 7.

We shall also need a partial converse to Proposition 8. Take a $g \in P[0, 2]$ and call it *nonsingular* (at zero) if its first Peano derivative at 0 is not zero and if for any $\epsilon > 0$ one has $\inf_x\{g(x); |x| \geq \epsilon\} > 0$.

Proposition 9. *If $g \in P[0,2]$ is nonsingular and the $X_{n,j} = g(Y_{n,j})$ satisfy conditions C_1 and C_2, so do the $Y_{n,j}$.*

This is obvious for C_1 and not difficult for C_2. Indeed, for y sufficiently small $|x| = |g(y)| \geq |ay|/2$, where a is the first derivative of g.

5.3 Limits for Binary Experiments

Binary experiment means an experiment $\mathcal{E} = \{\mathcal{P}_\theta; \theta \in \Theta\}$ where Θ is a set consisting of two points. These points will often be denoted 0 and 1, although other names will be used in examples.

For two probability measures P_0 and P_1, we shall use the following quantities:

(1) γ or $\gamma(P_0, P_1)$ is the affinity $\gamma = \int \sqrt{dP_0 dP_1}$,

(2) h is the Hellinger distance given by

$$h^2(P_0, P_1) = \frac{1}{2}\int (\sqrt{d}P_0 - \sqrt{d}P_1)^2 = 1 - \gamma,$$

(3) β is the mass of P_1 that is singular with respect to P_0.

Consider also the random variable

$$Y = \sqrt{\frac{dP_1}{dP_0}} - 1$$

with the distribution induced by P_0. The introduction of such variables may, at this point, seem entirely gratuitous. However, we shall soon see that their technical properties make them a very efficient vehicle to prove limit or approximation theorems.

One can then return to likelihood ratios or their logarithms later. To start, let us show that the first two moments of Y are linked to Hellinger distances. Indeed, the following relations are easily verifiable, with E denoting expectation under P_0:

(4) $EY = \gamma - 1 = -h^2,$

(5) $EY^2 = 2(1-\gamma) - \beta,$

(6) $\text{var}Y = (1-\gamma^2) - \beta.$

We have already noted (Section 3.2) that

$$h^2 \leq \frac{1}{2}\|P_0 - P_1\| \leq h\sqrt{2-h^2} \leq h\sqrt{2}.$$

Now consider the corresponding quantities and variables for experiments $\mathcal{E}_{n,j} = \{p_{0,n,j}, p_{1,n,j}\}$ so that h becomes $h_{n,j}$, Y becomes $Y_{n,j}$, and so on. We shall consider the direct product experiment $\mathcal{E}_n = \bigotimes_j \mathcal{E}_{n,j}$. This will be done under two conditions (B) and (N). The first is a boundedness assumption; the second is a negligibility assumption:

(B) There is a number $b < \infty$ such that $\sum_j h_{n,j}^2 \leq b$ for all n.

(N) $\lim_{n\to\infty} \sup_j h_{n,j}^2 = 0.$

Since the Hellinger affinity between the product measures $P_{0,n} = \prod_j p_{0,n,j}$ and $P_{1,n} = \prod_j p_{1,n,j}$ is the product $\prod_j (1 - h_{n,j}^2)$ of the individual affinities, condition (B) viewed under (N) just says that the two product measures $P_{0,n}$ and $P_{1,n}$ do not separate entirely. Indeed, if all the $h_{n,j}$ are in an interval $[0, h_0]$ with $h_0 < 1$ there is a c depending on h_0 for which

$$\exp(-ch_{n,j}^2) \leq 1 - h_{n,j}^2 \leq \exp(-h_{n,j}^2)$$

and similarly for the products. Condition (N) says that the $Y_{n,j}$ form an infinitesimal array, or in Loève's terminology, they are UAN.

The following three technical lemmas are simple but very important.

Lemma 1. *Let the conditions (B) and (N) be satisfied and let E denote expectation under $P_{0,n}$. Then*

$$(i) \quad -b \leq \sum_j EY_{n,j} \leq 0,$$

$$(ii) \quad \sum_j EY_{n,j}^2 \leq 2b,$$

$$(iii) \quad \sup_j EY_{n,j}^2 \to 0 \quad \text{as} \quad n \to \infty.$$

The sequence of distributions, under $P_{0,n}$, of $\sum_j Y_{n,j}$ is relatively compact on the line $(-\infty, +\infty)$.

Proof. The first three assertions are immediate consequences of the relations (4), (5), and (6) written at the beginning of this section. The relative compactness follows from the fact that the $\sum_j Y_{n,j}$ have bounded expectations and variances. □

Lemma 2. *Let (B) and (N) be satisfied and let A be an arbitrary subset of the line, subject only to the condition that 0 be an interior point of A. Let I_A denote the indicator of A. Furthermore, let $c_{n,j}$ be the conditional expectation*

$$c_{n,j} = \frac{EY_{n,j}I_A(Y_{n,j})}{EI_A(Y_{n,j})}.$$

Then $\sum_j c_{n,j}^2$ tends to zero as n tends to infinity.

Proof. By (iii) of Lemma 1, the denominator in $c_{n,j}$ can be assumed to be as close to unity as one pleases. Thus, the denominator can be ignored in the proof.

Note that under (B) and (N), the sum $\sum_j [EY_{n,j}]^2$ tends to zero. Indeed, it is equal to

$$\sum_j h_{n,j}^4 \leq [\sup_j h_{n,j}^2] \sum_j h_{n,j}^2.$$

To proceed further, let B denote the complement of A. Write

$$EY_{n,j}I_A(Y_{n,j}) = EY_{n,j} - EY_{n,j}I_B(Y_{n,j}).$$

This yields

$$\sum_j [EY_{n,j}I_A(Y_{n,j})]^2 \leq 2\sum_j [EY_{n,j}]^2 + 2\sum_j [EY_{n,j}I_B(Y_{n,j}]^2.$$

As observed earlier the first term on the right tends to zero. For the second take an interval $(-\tau, +\tau) \subset A$. Let $J_{n,j} = I[|Y_{n,j}| \geq \tau]$. Then

$$|EY_{n,j}I_B(Y_{n,j})| \leq E|Y_{n,j}|J_{n,j} \leq \frac{1}{\tau}EY_{n,j}^2 J_{n,j}.$$

Thus, $|EY_{n,j}I_B(Y_{n,j})|^2 \leq \frac{1}{\tau^2}|EY_{n,j}^2|^2$. However, $\sum_j |EY_{n,j}^2|^2 \leq 4\sum_j h_{n,j}^4 \to 0$. Hence the result as is claimed. □

If one takes a function of the $Y_{n,j}$ there is a similar result. Consider a function f of the class $P[0, 2]$. Near zero it has an expansion:

$$f(x) = ax + \frac{1}{2}bx^2 + x^2R(x).$$

If $b \neq 0$, there is a $\tau > 0$ such that for $|x| \leq \tau$ one has $|R(x)| < |b|/2$. Such a τ will occur below.

Lemma 3. *Let conditions (B) and (N) be satisfied. Let A be a set that contains 0 as interior point and let f be a measurable function of the class $P[0, 2]$ that remains bounded on A. Let $c_{n,j}$ be the conditional expectation*

$$c_{n,j} = \frac{Ef(Y_{n,j})I[Y_{n,j} \in A]}{EI[Y_{n,j} \in A]}.$$

Then $\sum_j c_{n,j}^2$ tends to zero as n tends to infinity.

Proof. Consider an interval $[-\tau, +\tau]$ contained in A and on which $|R(x)| \leq 1$ or $|R(x)| \leq |b|/2$ as mentioned before the statement of the lemma. The contribution to the $c_{n,j}$ of such an interval will involve terms of the type $aEY_{n,j}I[|Y_{n,j}| \leq \tau]$ and terms bounded by a multiple of $EY_{n,j}^2 I[|Y_{n,j}| \leq \tau]$. We already know by Lemma 2 that $\sum_j (EY_{n,j}^2 I[|Y_{n,j}| \leq \tau])^2$ tends to zero. It remains to take care of the contributions of the part, say, B_τ

of A situated outside the interval $[-\tau, +\tau]$. In B_τ, the function f is by assumption bounded by some number, say, C. It will contribute a term bounded in absolute value by something of the type $CPr[Y_{n,j} \in B_\tau]$. By Proposition 5 of Section 5.2.4, the sum $\sum_j Pr[Y_{n,j} \in B_\tau]$ stays bounded, thus $\sum_j (Pr[Y_{n,j} \in B_\tau])^2$ goes to zero. Hence the result. □

These lemmas about the negligibility of the squares of conditional expectations have a very pleasant implication about approximating the distributions of sums. To state one, let f be a function of the class $P[0,2]$. Assume that f is measurable and real-valued. If m is any positive measure, let $f(m)$ be its image by f. (This notation differs from the more common $m \circ f^{-1}$, but the common notation hides the fact that f carries measures in the same manner as it carries points. Our notation is also inspired by the fact that if m is the distribution of a random variable Y, the distribution of $f(Y)$ becomes $f(m)$.)

Theorem 1. *Let conditions (B) and (N) be satisfied. Let f be a measurable real-valued function of the class $P[0,2]$. Let $F_{n,j}$ be the distribution of $Y_{n,j}$ and let $L_n = \sum_j F_{n,j}$ be the Lévy measure with masses at zero deleted. Then the Lévy distance between the distribution of $\sum_j f(Y_{n,j})$ and the convolution exponential $\exp[M_n - \|M_n\|]$, where $M_n = f(L_n)$ is the image of L_n by f, tends to zero as n tends to infinity.*

Proof. Let

$$a_{n,j} = E[f(Y_{n,j}) | |Y_{n,j}| \leq \tau]$$

where τ is some arbitrary number such that f remains bounded on $[-\tau, +\tau]$. Let $Z_{n,j} = f(Y_{n,j}) - a_{n,j}$. According to the Lévy splittings in Section 5.2.2 and Proposition 6 of Section 5.2.5, we know that the distribution of $\sum_j Z_{n,j}$ can be approximated by an exponential in which the Lévy measure is $\sum_j \mathcal{L}(Z_{n,j})$. Let $A_n = \sum_j a_{n,j}$. Let ν_j be independent Poisson variables with expectations unity and let $Z_{n,j,k}$, $k = 1, 2, \ldots$ be independent replicates of $Z_{n,j}$.

This means that the distribution of $\sum_j f(Y_{n,j})$ is approximated in the Lévy distance by that of

$$A_n + \sum_j \sum_{k \leq \nu_j} Z_{n,j,k}.$$

Rewrite that last double sum as

$$\sum_j \sum_{k \leq \nu_j} f(Y_{n,j,k}) - \sum_j a_{n,j}\nu_j,$$

where the $Y_{n,j,k}$ are independent replicates of the $Y_{n,j}$.

To obtain the result, it is enough to show that $\sum_j a_{n,j}\nu_j$ and $A_n = \sum_j a_{n,j}$ differ little in probability. However $\sum_j a_{n,j}\nu_j$ has for expectation A_n. By Lemma 3, its variance $\sum_j a^2_{n,j}$ tends to zero. The desired result follows by Chebyshev's inequality. □

Remark 2. We have assumed that f is real-valued. This would exclude the function $y \rightsquigarrow 2\log(1+y)$ defined on $[-1, \infty)$ that gives the log likelihoods and would be unfortunate. However, note that convolution of finite signed measures on the additive semigroup $[-\infty, +\infty)$ is well defined, and so are sums of random variables with values in that semigroup. The entire proof would remain valid, but one has to redefine the Lévy metric or analogies to accommodate the point $-\infty$. One possibility is to define a distance on $[-\infty, +\infty)$ as follows. If $x_i, i = 1, 2$ are two points of $[-\infty, \infty)$, let their distance be $|\exp(x_1) - \exp(x_2)|$. Then define the distance between measures by the corresponding dual Lipschitz norm. With such definitions the theorem remains entirely correct with the same proof.

Remark 3. We have used implicitly the fact that conditions (B) and (N) imply that the sum $\sum_j f(Y_{n,j})$ is shift-compact, because we have relied on Proposition 6, Section 5.2.5, for the replacement of a convolution product by its accompanying exponential. This could have been avoided by using the more general result given in Remark 1 following the proof of Proposition 6 in Section 5.2.5. To deduce the desired exponential approximation for sums such as $\sum_j f(Y_{n,j})$ from conditions on the $Y_{n.j}$, one can use the properties of the Lévy splittings. Take a τ as in the proof of Theorem 1 and define the $a_{n,j}$ accordingly as done earlier. The Lévy splitting technique then gives a representation

$$\begin{aligned} f(Y_{n,j}) - a_{n,j} &= (1-\eta_{n,j})U_{n,j} + \xi_{n,j}V_{n,j} \\ &+ (\eta_{n,j} - \xi_{n,j})U_{n,j}. \end{aligned}$$

It is not difficult to see that, in the sum of such expressions, the sum $\sum_j(\eta_{n,j}-\xi_{n,j})U_{n,j}$ is always negligible compared to the rest. A precise expression of that fact occurs in Le Cam [1986, chapter 15, section 2, proposition 3]. Another form would use lemma 2 of Le Cam's chapter 15, section 2. Informally and roughly the sum $\sum_j(1-\eta_{n,j})U_{n,j}$ being of either small variance or not much more concentrated than a normal with variance $(1-\alpha)\sigma^2$ has a distribution that is not changed much by addition of a term $\sum_j(\eta_{n,j}-\xi_{n,j})U_{n,j}$ that has mean 0 and variance $2\alpha(1-\alpha)\sigma^2$. This removes one place where some shift-compactness was used in the proof of Proposition 6, Section 5.2.5.

Thus, it is enough to argue about the sum of the two *independent* terms $\sum_j(1-\eta_{n,j})U_{n,j}$ and $\sum_j \xi_{n,j}V_{n,j}$. For the first, the exponential approximation is always valid, either in Lévy distance if the variance of the sum is small or in Kolmogorov distance if the variance is large, as results from the Lindeberg argument.

For second term, the proof of Proposition 6, Section 5.2.5, uses that $\sum Pr[\xi_{n,j}V_{n,j} \neq 0]$ stays bounded. This will be true here of the $f(Y_{n,j})$ whenever it is true of the $Y_{n,j}$ themselves. Hence the result.

There is a partial converse: Suppose that f is as stated in the theorem. Suppose also that f is *nonsingular at zero, increasing to infinity as y increases to infinity.* Then, under *(N)*, shift-compactness of $\sum_j f(Y_{n,j})$ implies shift-compactness of $\sum_j Y_{n,j}$. This is an easy consequence of the preceding and Proposition 5 of Section 5.2.4.

To illustrate, let us consider the loglikelihoods $\Lambda_{n,j}$ of the experiment $\mathcal{E}_{n,j}$. They can be written as $\Lambda_{n,j} = f(Y_{n,j})$ with $f(y) = 2\log(1+y)$. The sum $\sum_j \Lambda_{n,j}$ is the loglikelihood of the product measures $\Lambda_n = \log(dP_{1,n}/dP_{0,n})$.

According to Theorem 1, the distribution of Λ_n is approximated by the Poisson exponential $\exp[M_n - \|M_n\|]$, where M_n is the image $M_n = f(L_n)$. Taking into account the linearity of the transformation $\mu \rightsquigarrow f(\mu)$ as applied to measures, it is easy to verify that this is exactly the approximation yielded by the poissonization described for experiments in Section 4.3. (The linearity of $\mu \rightsquigarrow f(\mu)$ is more visible in the notation $\mu \circ f^{-1}$.) The fact that the two approximations coincide is reassuring. It also comes with a slight element of surprise. After all, the arguments described in Section 5.2 were of a general nature and not designed to preserve the status of the variables as issued from likelihood ratios in experiments.

Theorem 1 indicates that one can obtain approximations to the distributions of sums $\sum_j f(Y_{n,j})$ from knowledge about the distributions of $\sum_j Y_{n,j}$. It is sometimes more convenient to use other variables, for example, the variables $f(Y_{n,j} = (1+Y_{n,j})^2 - 1 = (dp_{1,n,j}/dp_{0,n,j}) - 1$. Those turn out to be convenient if the numerator measure is a mixture; see Le Cam and Yang [1988].

Remark 4. Generally, under condition (N), if f is a measurable function of class $P[0,2]$ one-to-one, and nonsingular at zero, from approximations to the distributions of $\sum_j f(Y_{n,j})$ one can deduce approximations to the distributions of $\sum_j Y_{n,j}$.

So far we have considered *approximations.* Now let us consider *limits* of distributions for the Λ_n and other variables. But first look at formulas for Hellinger transforms. The Hellinger transform of our product experiment $\mathcal{E}_n$ is

$$\int (dP_{0,n})^{1-z}(dP_{1,n})^z = \prod_j \int (dp_{0,n,j})^{1-z}(dp_{1,n,j})^z.$$

If one replaces $\mathcal{E}_n$ by its accompanying poissonized experiment, one obtains a Hellinger transform whose logarithm, for $0 < z < 1$, is given by

$$\psi_n = \int [e^{z\lambda} - 1] M_n(d\lambda) = \int [(1+y)^{2z} - 1] L_n(dy).$$

See the formulas in Chapter 3. Note that the last formula on the right is *not* an approximation to a logarithm of Laplace transform or characteristic function for $\sum_j Y_{n,j}$. For the latter we shall take the logarithm of the characteristic function of the poissonized sum of the $Y_{n,j}$. It is given by the integral

$$\int (e^{ity} - 1) L_n(dy) \quad \text{for } t \text{ real}$$

and will intervene only in the characterization of limits.

Assume that conditions (B) and (N) are satisfied and let F_n be the law of $\sum_j Y_{n,j}$. The F_n are relatively compact on the line. Taking a subsequence if necessary, assume that the F_n converge to a limit F. The corresponding Lévy measures L_n do not necessarily form a relatively compact sequence for the ordinary convergence because, in particular they might not be bounded in total mass. However, the Khinchin measures K_n defined by

$$K_n(dy) = [\min(1, y^2)] L_n(dy)$$

form a relatively compact sequence. This is true here without "recentering" the $Y_{n,j}$ because the integrals $\int y^2 L_n(dy) \leq 2 \sum_j h_{n,j}^2$ remain bounded. Before going any further, let us state a simple but useful fact.

Lemma 4. *Let* $a_n = \sum_j h_{n,j}^2$ *with* $h_{n,j}^2 = EY_{n,j}$. *Then* $a_n = -\int y F_n(dy) = -\int y L_n(dy)$. *If the* F_n *converge weakly to a limit* F *then the* a_n *converge to a limit* a *and* $a = -\int y F(dy)$.

Proof. This is an easy consequence of the fact that the second moments of the F_n stay bounded. Indeed, a bit of algebra shows that $\int y^2 F_n(dy) \leq a_n^2 + \operatorname{var}(\sum_j Y_{n,j}) \leq a_n(2 + a_n)$. □

These results, and especially Theorem 1, give a handle in passing from sums of the type $\sum_j Y_{n,j}$ to sums of logarithms of likelihood ratios, but this is in terms of the Lévy measures of the accompanying poissonized versions. It does not actually imply statements in which just knowing the form of a limit such as $F = \lim_n F_n$ one can automatically obtain the limits of distributions of other sums of the type $\sum_j f(Y_{n,j})$ because we have not yet proved the appropriate uniqueness theorems.

Here is such a uniqueness result. It is stated in terms of Fourier transforms to avoid a discussion of the proper definition of logarithms in the convolution algebra of measures. It also involves the choice of a function u defined on the line such that $0 \leq u(x) \leq 1$ for all x. We also assume that u is continuous, equal to unity in a neighborhood of zero (for instance $(-1/2, +1/2)$), and that it vanishes outside $[-1, +1]$.

Theorem 2. *Let H be an infinitely divisible distribution on the real line. Let $\phi(t) = \int e^{itx} H(dx)$ be its characteristic function. Then*

(i) $\phi(t)$ never vanishes and there is a unique function $t \rightsquigarrow \psi(t)$ such that $\psi(0) = 0$, ψ is continuous in t, and $\phi(t) = \exp[\psi(t)]$ for all $t \in (-\infty, \infty)$.

(ii) ψ can be written as

$$\psi(t) = imt + \int [e^{itx} - 1 - itxu(x)][1 \wedge x^2]^{-1} K(dx),$$

where K is a finite positive measure called the Khinchin measure and m is a real number.

(iii) The pair (m, K) is uniquely defined.

(iv) If H_n is a sequence of infinitely divisible distributions corresponding to pairs (m_n, K_n), convergence of H_n to H is equivalent to convergence of (m_n, K_n) to (m, K).

Note 4. In part (ii) of the theorem, the integrand is taken to be $-t^2/2$ at $x = 0$. The convergences in (iv) are in the ordinary sense.

Proof. (i) The fact that ϕ does not vanish is simple: For each integer k one can write $\phi(t) = [\phi_k(t)]^k$ for some characteristic function ϕ_k. As k tends to infinity $\phi_k(t)$ tends to unity for each t. That is impossible if $\phi(t) = 0$. The possibility of defining a *continuous* logarithm follows from that and the continuity of ϕ.

(ii) The existence of the representation in (ii) is obtainable by passage to the limit in the poissonization of the power $[\phi_k]^k$.

(iii) To prove the uniqueness, take the second difference

$$\begin{aligned} \omega_s(t) &= \psi(t) - \frac{1}{2}[\psi(t+s) + \psi(t-s)] \\ &= \int e^{itx}(1 - \cos sx)[1 \wedge x^2]^{-1} K(dx). \end{aligned}$$

Introducing a function $l(s,x) = (1 - \cos(sx))[1 \wedge x^2]^{-1}$, the last integral may be written

$$\omega_s(t) = \int e^{itx} l(s,x) K(dx).$$

Thus, by the uniqueness of Fourier transforms, the measures $l(s,x)K(dx)$ are well determined. Since $l(s,x)$ cannot vanish at x for two irrationally related values of s, the measure K is determined. It follows that the constant m is also determined.

(iv) The equivalence of convergences can be seen as follows. First assume that the pairs $(m_n.K_n)$ converge. Then so do the corresponding characteristic functions. Hence the measures H_n. Conversely, assume that H_n tends to H. Then by Proposition 5 of Section 2, the Khinchin measures must form a relatively compact sequence. Therefore, taking a subsequence if necessary, one can assume that they converge to some limit, say, K'. Let K be the Khinchin measure corresponding to H. The argument used to prove (iii) shows that K and K' must be the same. The convergence of the constants m_n follows immediately. This concludes the proof of the theorem. □

Remark 5. The uniqueness statement might puzzle the reader, since the constants called m obviously can depend on the choice of the *arbitrary* function u. We assume that a choice of u has been made. The introduction of the term $u(x)itx$ is to ensure convergence of the integrals, or in Lévy's derivations, to compensate the small jumps. The way to achieve this varies with the authors. For example, instead of our $xu(x)$, Khinchin and Lévy use $x/(1+x^2)$. Feller uses $\sin x$. It is quite immaterial but changes the m.

Remark 6. In Theorem 2, we used the Khinchin measures K_n instead of Lévy measures. This is simply because convergence properties are easier to state in terms of Khinchin measures. However, when using transformed variables such as $f(Y_{n,j})$ instead of $Y_{n,j}$, the Lévy measures are easier to handle. They also have a clearer interpretation. The proof of our next assertion uses still another possibility. The theorem gives conditions for contiguity in terms of the limits of the distributions F_n of the sums $\sum_j Y_{n,j}$.

Consider a finite set of experiments $\mathcal{E}_{n,j} = \{p_{0,n,j}, p_{1,n,j}\}$ and their product, $\mathcal{E}_n = \{P_{0,n}, P_{1,n}\}$. Recall that $\beta_{n,j}$ is the mass of the part of $p_{1,n,j}$ that is $p_{o,n,j}$ singular.

Let L be the limit of the Lévy measures L_n as defined in Theorem 1.

Theorem 3. *Assume that conditions (B) and (N) hold. The two sequences $\{P_{0,n}\}$ and $\{P_{1,n}\}$ are contiguous if and only if the following three conditions hold:*

(i) $L[\{-1\}] = 0$,

(ii) $\sum_j \beta_{n,j} \to 0$ *as* $n \to \infty$,

(iii) the variance of F is the limit of the variances of the F_n.

Proof. For $z \in (0,1)$ open, the expression $\int[(1+y)^{2z} - 1]L_n(dy)$ is an approximation to the logarithm of Hellinger transform

$$\psi_n(x) = \log \int (dP_{0,n})^{1-z}(dP_{1,n})^z,$$

as follows from Theorem 1 and the fact that the Hellinger transforms are bounded away from zero by condition (B).

To pass to the limit in the expression $\int[(1+y)^{2z} - 1]L_n(dy)$, rewrite it as

$$2z \int yL_n(dy) + \int [(1+y)^{2z} - 1 - 2zy]L_n(dy).$$

Note that for y small $[(1+y)^{2z}-1-2zy] = 2z(2z-1)\frac{y^2}{2}+\ldots$ It follows that

$$\lim_{\tau\to 0}\lim_{n\to\infty}\int_{|y|<\tau}[(1+y)^{2z}-1-2zy]L_n(dy) = 2z(2z-1)\frac{\sigma^2}{2},$$

where σ^2 is the variance of the gaussian component of the limit F. Recall that $\int yF_n dy \to a = \int yF dy$. Accordingly the ψ_n converge on $(0,1)$ to a limit ψ such that

$$\psi(z) = -(2a+\sigma^2)z + 2\sigma^2 z^2 + \int[(1+y)^{2z}-1-2zy]L(dy).$$

According to Proposition 1, Chapter 3, the contiguity of $\{P_{0,n}\}$ to $\{P_{1,n}\}$ is equivalent to $\lim_{z\to 0, z>0}\psi(z) = 0$. This gives that

$$-L[\{-1\}] = \lim_{z\to 0, z>0}\psi(z) = 0.$$

Similarly, contiguity of $\{P_{1,n}\}$ to $\{P_{0,n}\}$ is equivalent to

$$\lim_{z\to 1, z<1}\psi(z) = -2a+\sigma^2+\int y^2 L(dy) = 0.$$

Call s^2 the variance of the limiting distribution F. It equals $s^2 = \sigma^2 + \int y^2 L(dy)$ and is always at most equal to the limit of the variances of the F_n. This gives

$$\lim_n \sum_j \left[h_{n,j}^2(2-h_{n,j}^2) - \beta_{n,j}\right] = 2a - \lim_n \sum \beta_{n,j} = s^2.$$

Thus $2a - s^2 \geq \lim_n \sum \beta_{n,j}$. However, the contiguity condition also gives

$$0 = 2a - s^2 = 2a - [\sigma^2 + \int y^2 L(dy)].$$

This implies $\lim_n \sum \beta_{n,j} = 0$ and that the variance $\sigma^2 + \int y^2 L(dy)$ is the limit $2a$ of the variances of the F_n. □

Note 5. The asymmetry in the conditions $L[\{-1\}] = 0$ and the pair $\lim_n \sum_j \beta_{n,j} = 0$ and variance of $F =$ limit of the variances of F_n is due to the use of the ratios

$$\frac{dp_{1,n,j}}{dp_{0,n,j}} = X_{n,j} + 1$$

with the choice of $P_{0,n}$ in the denominator. Here $X_{n,j} = (Y_{n,j} + 1)^2 - 1$.

One could also rephrase the condition in terms of the limit of the distribution of $\sum_j X_{n,j} = \sum_j [(1 + Y_{n,j})^2 - 1]$.

Then, the condition that the variance of F is the limit of the variances of the F_n becomes the condition that the expectation of the limit of the distributions of the $\sum_j X_{n,j}$ is the limit of the expectations of $\sum_j X_{n,j}$.

5.4 Gaussian Limits

In the preceding sections we have described possible limits under conditions (B) and (N) for $\sum_j Y_{n,j}$ and consequently $\sum_j \Lambda_{n,j}$ where the $\Lambda_{n,j}$ are the logarithms of likelihood ratios. This was done for the case of binary experiments, but extension to the case of experiments indexed by a set Θ of finite cardinality is not difficult.

Here we shall concentrate on the special case of gaussian limits, because of their importance and peculiarities. The begining of the section is about binary experiments, but more general index sets will also be discussed. Besides our usual $Y_{n,j}$ and $\Lambda_{n,j}$ the variables $X_{n,j} = (1 + Y_{n,j})^2 - 1 = (dp_{1,n,j}/dp_{0,n,j}) - 1$ will play an important role. It will be convenient to use the following language.

Definition 1. *A sequence of distributions or random variables, such as $\sum_j Y_{n,j}$, is called asymptotically normal (or gaussian) if every subsequence contains a further subsequence tending to a gaussian limit.*

Definition 2. *A sequence of random variables is asymptotically degenerate if every subsequence contains a further subsequence tending to a nonrandom limit.*

The existence of limits implies relative compactness. Later we shall mention cases where relative compactness does not hold.

In the binary case, the following result is very useful. All distributions are under the measures $P_{0,n}$.

Theorem 4. *Let the conditions (B) and (N) be satisfied. Then the following statements are all equivalent:*

(i) the variables $\sum_j Y_{n,j}$ are asymptotically normal;

(ii) the variables $\sum_j X_{n,j}$ are asymptotically normal;

(iii) the variables $\sum \Lambda_{n,j}$ are asymptotically normal;

(iv) the variables $\sum_j(\Lambda_{n,j} - 2Y_{n,j})$ are asymptotically degenerate.

If these conditions are satisfied then $s_n^2 + \sum_j(\Lambda_{n,j} - 2Y_{n,j})$ tends to zero in $P_{0,n}$ probability for numbers $s_n^2 = \sum_j s_{n,j}^2$ with $s_{n,j}^2 = EY_{n,j}^2 I[|Y_{n,j}| \leq \tau]$ where $\tau > 0$ is arbitrary.

Proof. Conditions (B) and (N) imply that the variables under consideration form infinitesimal arrays. The conditions, together with our previous remarks on $P[0,2]$ functions, also imply the relative compactness of the sums in (i) to (iv). According to Section 2, these sums will be asymptotically normal if and only if for every $\epsilon > 0$ the sums $\sum_j Pr[|Y_{n,j}| > \epsilon]$ tend to zero. This is clear for the $Y_{n,j}$ and follows for the other variables by simple computation. For example, in statement (iv), one can use the fact that for every ϵ there is a δ such that $|\log(1+y) - y| > \delta$ means the same as $|y| > \epsilon$.

Now let us go back to (iv). By Peano differentiability of $\log(1+y)$

$$\sum_j [\log(1 + Y_{n,j}) - Y_{n,j} + \frac{1}{2}Y_{n,j}^2] \to 0$$

in probability. Thus our sum in (iv) behaves like $-\sum_j Y_{n,j}^2$, whenever the asymptotic normality in any one of (i) to (iv) holds. Consider then the variables

$$W_{n,j} = Y_{n,j}^2 I[|Y_{n,j}| \leq \tau] \quad \text{for a } \tau > 0.$$

The behavior of $\sum_j W_{n,j}$ as $n \to \infty$ will not depend on the choice of τ, because for every $\epsilon > 0$ the sum $\sum_j Pr[W_{n,j} > \epsilon]$ tends to zero. The same remark applies to the variance of $\sum_j W_{n,j}$.

This variance is arbitrarily small, because it does not exceed $2\tau^2 \sum_j h_{n,j}^2$ and τ can be arbitrarily small. In summary, $\sum_j W_{n,j}$ behaves like its expectation, as stated in (iv). We already noted that (iv), under (B) and (N), implies the asymptotic normal behavior. Thus, the theorem is proved. □

Remark 7. For practical purposes the statement of the theorem should be complemented as follows. Suppose that the distribution of $\sum_j Y_{n,j}$ is approximated by the normal $\mathcal{N}(m_n, \sigma_n^2)$ of mean m_n and variance σ_n^2. According to Lemma 2, Section 5.3, one can replace σ_n^2 by the s_n^2 defined in the statement of the theorem, that is, $s_n^2 = \sum_j EY_{n,j}^2 I[|Y_{n,j}| \leq \tau]$. Then the distribution of $\sum_j X_{n,j}$ is approximated by $\mathcal{N}(2m_n + s_n^2, 4s_n^2)$ and that of $\sum_j \Lambda_{n,j}$ is approximated by $\mathcal{N}(2m_n - s_n^2, 4s_n^2)$. This follows from statement (iv) in the theorem and the remark that the final part of the proof also applies to the variables $X_{n,j} = (1 + Y_{n,j})^2 - 1 = 2Y_{n,j} + Y_{n,j}^2$.

Remark 8. The relation between the asymptotic degeneracy of $\sum_j Y_{n,j}^2$ and the asymptotic normality of $\sum_j Y_{n,j}$ is a a classical result attributed to Bobrov [1937]. A far-reaching extension occurs in martingale theory; see McLeish [1975] and Rebolledo [1980]. Note that what enters into the arguments are expectations and variances of *truncated* variables.

The preceding theorem does not say anything about contiguity. However, it is clear that asymptotic normality of the log-likelihood $\log(dP_{1,n}/dP_{0,n}) = \sum_j \Lambda_{n,j}$ always implies contiguity of $\{P_{0,n}\}$ to $\{P_{1,n}\}$. This is because, in the notation of Chapter 3, Theorem 1, the condition $M[\{-\infty\}] = 0$ is satisfied. Here, the contiguity condition of $\{P_{1,n}\}$ to $\{P_{0,n}\}$ takes the following form.

Proposition 10. *Let $\{P_{0,n}\}$ and $\{P_{1,n}\}$ be two sequences of probability measures. Let $\Lambda_n = \log(dP_{1,n}/dP_{0,n})$ be considered as a random element with the distribution induced by $P_{0,n}$. Assume that the distributions of Λ_n converge to a normal limit $\mathcal{N}(\mu, \sigma^2)$. Then the sequences $\{P_{0,n}\}$ and $\{P_{1,n}\}$ are contiguous if and only if $\mu = -\sigma^2/2$.*

This is an immediate consequence of Theorem 1, Chapter 3. Under the conditions of Theorem 1 and assuming asymptotic normality, the contiguity means that, in the notation of Remark 2, $m_n + (1/2)s_n^2$ tends to zero. In the present case we know of a good approximation for m_n. Indeed, according to Section 5.3, Lemma 4, one can take $m_n = -a_n$ with $a_n = \sum_j h_{n,j}^2$. Now the number s_n^2 was the sum of second moments of truncated variables $Y_{n,j}I[|Y_{n,j}| \leq \tau]$. The sum of the untruncated second moments is

$$\sum_j EY_{n,j}^2 = 2a_n - \sum_j \beta_{n,j}.$$

Therefore, the convergence to zero of $m_n + (1/2)s_n^2$ implies that $\sum_j \beta_{n,j} \to 0$ and $\sum_j EY_{n,j}^2 I[|Y_{n,j}| > \tau]$ tends to zero as $n \to \infty$. This last condition is a condition of the Lindeberg type. We say of "Lindeberg type" because statements of Lindeberg's condition vary from text to text, depending on whether the variables have been centered on the expectations of truncated variables. Here this does not matter because of Lemma 2, Section 3.

The Lindeberg condition was intended as a necessary and sufficient condition for asymptotic normality of sums *with preservation of variances in the passage to the limit.* It is not necessary for asymptotic normality per se. It can be satisfied by the $Y_{n,j}$ and not by the $X_{n,j}$, for instance.

To elaborate a bit further, note that this preservation of variances does not by itself imply that the sum of singular parts $\sum_j \beta_{n,j}$ tends to zero. An example can be constructed by taking $p_{0,n,j}$ to be the Lebesgue measure on $[0, 1]$ and taking $p_{1,n,j}$ uniform on $[0, (1-(1/n))^{-2}]$ for $j = 1, \ldots, n$. Here our $Y_{n,j}$ are identically $-1/n$. Their sum is identically -1 and the variances are zero. They are certainly preserved in the passages to the limit. But $\sum_j \beta_{n,j}$ does not tend to zero. This shows that in the normal approximations to the distributions of $\sum_j Y_{n,j}$ or $\sum_j \Lambda_{n,j}$ the fact that expectations are close to minus half the variances is a strong condition. A similar remark applies to the normal approximations to the law of $\sum_j X_{n,j}$. Contiguity requires that expectations be close to zero, not only that they be preserved by passage to the limit. This fact is often forgotten because authors readily assume, sometimes without explicit-mention, that the $p_{0,n,j}$ dominate the $p_{1,n,j}$.

So far, we have considered only the case where the parameter set Θ has only two elements. In the asymptotically normal situation there is a simple remark that allows passages from two-element sets to more general ones. It is as follows.

Consider random elements Z_n with values in R^k. Let $Z_n^{(i)}$ be the ith coordinate of Z_n. Call the sequence $\{Z_n\}$ asymptotically infinitely divisible if every subsequence contains a further subsequence whose law tends to an infinitely divisible one.

Proposition 11. *Assume that $\{Z_n\}$ is asymptotically infinitely divisible and that each of the coordinate sequences $\{Z_n^{(i)}\}$ is asymptotically normal. Then the k-dimensional (joint) distribution of Z_n is asymptotically normal.*

Proof. A proof can be constructed using the k-dimensional analogue of the Khinchin-Lévy representation for infinitely divisible distributions. Taking subsequences as necessary, assume that the distributions of the Z_n converge. The limit corresponds to a Khinchin measure K. The asymptotic normality of the coordinate $Z_n^{(i)}$ means that the projection of K on the ith axis is concentrated at zero. This being true for every i, the measure K itself must be concentrated at the origin. □

The result admits various extensions. If k is finite, tightness of the coordinate distributions implies tightness of the joint distributions. This is no longer the case in infinite-dimensional vector spaces. However, the result as stated remains entirely valid for infinitely dimensional processes $Z_n^\theta, \theta \in \Theta$, for the supremum norm. For the finite-dimensional case, there are also extensions that do not require the relative compactness implied in the statement of the proposition. One would, for instance, replace the asymptotic infinite divisibility by approximability of the joint distributions by infinitely divisible ones in the sense of a distance defined analogously to the Lévy distance using, instead of intervals, hyperrectangles with sides parallel to the coordinate axes.

One of the consequences of this state of affairs can be described in the context of the product $\mathcal{E}_n = \bigotimes_j \mathcal{E}_{n,j}$ with $\mathcal{E}_{n,j} = \{p_{t,n,j}; t \in \Theta\}$ mentioned in the introduction to this chapter. To this effect, suppose that Θ is a finite set, $\Theta = \{0, 1, \ldots, k\}$.

Proposition 12. *Assume that Θ is finite and that for each $t \in \Theta$ the experiments $\{p_{0,n,j}, p_{t,n,j}\}$ satisfy conditions (B) and (N). Suppose also that for each $t \in \Theta$ the binary experiments $\{P_{0,n}, P_{t,n}\}$ formed by the product measures are asymptotically gaussian.*

Then the experiments $\mathcal{E}_n$ are asymptotically gaussian.

Proof. By *asymptotically gaussian* we mean that there are gaussian experiments $\mathcal{G}_n$ such that the distance $\Delta(\mathcal{E}_n, \mathcal{G}_n)$ of Chapter 2 tends to zero.

Consider then $\Lambda_n^{(t)} = \log(dP_{t,n}/dP_{0,n})$. Condition (B), together with the approximability of the binary experiments by gaussian ones, implies contiguity of the sequences $P_{0,n}$ and $P_{t,n}$. According to Proposition 3, we also have *joint* asymptotic normality of the $\Lambda_n^{(t)}$. It follows then from a result of V. Strassen [1965], see also Le Cam [1986, p. 288, lemma 4] that there are small, possibly random, distortions of the vector Λ_n that are *exactly* gaussian. Let W_n be such a distortion. One can assume that it satisfies the relations $EW_n^{(t)} = -$ variance $W_n^{(t)}/2$. Then define $dQ_{t,n} = \exp[W_n^{(t)}]dP_{0,n}$. This provides the approximating gaussian experiment.

There are other facts that are expressions of the idea that gaussian experiments are some "degenerate" form of infinitely divisible ones. Such a fact occurs in the paper by Le Cam and Yang [1988]. That paper relies on conditions (B) and (N) and on a different expression of the conditions for asymptotic normality with contiguity. Consider first two probability measures P and Q on the same σ-field. Let $Z = dQ/(dP+dQ)$. Take a $\tau \in (0, 1)$. It is easy to see that

$$Q[\frac{Z}{1-Z} \geq 1+\tau] \geq \|Q - [Q \wedge (1+\tau)P\|.$$

Also, a simple computation shows that

$$\begin{aligned} \|Q - [Q \wedge (1+\tau)P\| \geq \; & \max\{\tau P[\frac{Z}{1-Z} > 1 + 2\tau], \\ & \tau Q[\frac{Z}{1-Z} > \frac{1+\tau}{1-\tau}]\} \end{aligned}$$

Call an experiment $\mathcal{F}$ weaker than $\mathcal{E}$ (in the sense of Blackwell) if, for any given loss function, every risk function for $\mathcal{F}$ can be matched by a risk function for $\mathcal{E}$ that is pointwise no larger.

Theorem 5. *Let Θ be a finite set and let $\mathcal{E}_{n,j}$ and $\mathcal{F}_{n,j}$ be experiments indexed by Θ. Let $\mathcal{E}_n$ be the direct product of the $\mathcal{E}_{n,j}$ and similarly for $\mathcal{F}_n$.*

Assume that each $\mathcal{F}_{n,j}$ is weaker than the corresponding $\mathcal{E}_{n,j}$. Assume that the $\mathcal{E}_{n,j}$ satisfy conditions (B) and (N) and that the $\mathcal{E}_n$ are asymptotically gaussian.

Then the $\mathcal{F}_{n,j}$ satisfy (B) and (N) and their products $\mathcal{F}_n$ are asymptotically gaussian.

Proof. The Hellinger distances relative to the $\mathcal{F}_{n,j}$ are smaller than those relative to the $\mathcal{E}_{n,j}$. This implies that Conditions (B) and (N) carry over to the weaker experiments. Similarly, let s and t be two points of Θ. Let $p_{s,n,j}$ be the measure corresponding to s in $\mathcal{E}_{n,j}$ and let $q_{s,n,j}$ be the corresponding measure for $\mathcal{F}_{n,j}$. By Theorem 2 of Section 3.2, there is a transition T (or Markov kernel) such that $q_{\theta,n.j} = Tp_{\theta,n.j}$ for all $\theta \in \Theta$. The linearity and positivity of T imply that

$$Tp_{s,n,j} \wedge (1+\tau)Tp_{t,n,j} \geq T[p_{s,n,j} \wedge (1+\tau)p_{t,n,j}].$$

According to the elaboration given just before the statement of the theorem, this implies that if the $\mathcal{E}_{n,j}$ satisfy the conditions of being asymptotically gaussian, so do the $\mathcal{F}_{n,j}$.

This result, together with the preservation of the LAN conditions to be described in the next chapter, has multiple applications. The $\mathcal{F}_{n,j}$ might be obtained from the $\mathcal{E}_{n,j}$ by loss of a variable, by truncation or censoring, by grouping, or by any one of various procedures that lead to a loss of "information." See Le Cam and Yang [1988], for many examples.

The reader may have noticed that we have covered only cases subject to conditions (B) and (N). This is sufficient for many classical studies on the "local" behavior of experiments. To pass to global properties, as exemplified by the global gaussian approximability of Nussbaum [1996], one needs more elaborate arguments and to quilt pieces together.

5.5 Historical Remarks

The study of independent observations has been the mainstay of statistics for a long time, going back at least to Laplace. In fact, a look at textbooks shows that the bulk of the derivations were until recently about independent identically distributed variables, with some forays into time series. In this respect, the present chapter remains in the historical tradition: It is about independent observations. More modern treatments rely on martingale theory and the remark by Doob [1953, p. 348] that for any sequence of observations, $X_1, \ldots, X_m, \ldots$ the successive likelihood ratios dQ_m/dP_m of the joint distributions of the $X_1, \ldots, X_m$ form a martingale or at least a lower semimartingale. This has been exploited by many authors. It does lead to widely applicable results, which, however, are not as precise or simple as those available in the independent case. See the book by Jacod and Shiryaev [1987].

The section on auxiliary results recalls some facts established in probability in the 1930s mostly by Khinchin and Paul Lévy. Of course, Lindeberg's method dates back to around 1920. For infinitesimal arrays, the fact that $\sup |X_{n,j}|$ must be small so that the sum will be approximately normal was known to Paul Lévy and Khinchin in the early 1930s. The published proofs tend to rely heavily on characteristic functions. Here we have attempted to minimize the use of Fourier or Laplace transforms. They do occur, but only for simplicity's sake.

A proof that does not rely on Fourier transforms occurs in a paper of Jeganathan and Le Cam [1985]. That paper was written to prove an assertion of Lévy on martingales with symmetric increments.

For sums of independent variables, Lévy was after a much more general theorem that did not assume the arrays to be infinitesimal. The solution uses a theorem of Cramér: If X and Y are independent, and the sum $X+Y$ is gaussian, then X and Y are gaussian. No "elementary" proof of this fact is known.

The results concerning Peano differentiable functions must have been known to K. Itô around 1940.

Section 5.3 on limit laws originated in handwritten notes of Le Cam, around 1961. These notes were communicated to J. Hájek and partially reproduced in Hájek and Šidák [1967], as well as in Le Cam [1966]. The passage from sums $\sum_j Y_{n,j}$ to sums of the form $\sum_j f(Y_{n,j})$ is described in a paper of Loève [1957]. However, Loève's variables had fairly arbitrary truncated expectations, not subject to the negligibility of Section 5.3, Lemma 2. This negligibility simplifies many statements.

Kolmogorov [1956] was interested in the approximation of laws of sums of independent variables by infinitely divisible ones. Here it is more natural to use the "poissonized" approximation of Section 5.3, Theorem 1. For a discussion of the subject and very sharp bounds, see the monograph by Arak and Zaitsev [1988].

The main facts of Section 5.4 are taken from Le Cam's notes of 1961 and the paper by Le Cam and Yang [1988]. The equivalence of convergence to normal limit for $\sum_j Y_{n,j}$ and convergence to a constant for $\sum_j Y_{n,j}^2$ occurs in results of Bobrov [1937], Raikov [1938], and Gnedenko [1939]. Propositions 2 and 3, Section 5.4, are mentioned in Le Cam [1986]. The Le Cam–Yang paper of 1988 originated in a study of G. Yang about Markov process models meant to describe the behavior of sodium channels in nerves. The Markov processes may have many states falling into two groups. In the first group, the channel registers as "open." In the second group, the channel registers "closed." One form of loss of information is due to this grouping of states; another form is due to the fact that several channels may be under the measuring instrument at once. Without loss of information, an asymptotic treatment was easy along classical lines.

The theorem says that asymptotic normality is preserved. We shall see later that the LAN conditions are also preserved, so that a straightforward treatment is available.

6

Local Asymptotic Normality

6.1 Introduction

The classical theory of asymptotics in statistics relies heavily on certain local quadratic approximations to the logarithms of likelihood ratios. Such approximations will be studied here but in a restricted framework.

We shall use families of probability measures $\{P_{\theta,n}; \theta \in \Theta\}$ indexed by a set Θ assumed to be either all of a k-dimensional euclidean space $\Re^k$ or an open subset of such a space. One can deal with more general situations, and one must do so for practical purposes. However, the results are more easily described in a restricted situation. They can then be generalized without much trouble.

The approximations in question are of a local character. The word "local" is meant to indicate that one looks at parameter values η so close to a point θ_0 that it is not possible to separate $P_{\theta_0,n}$ and $P_{\eta,n}$ easily. This can be made precise in various ways. However, to make the situation as simple as possible at the start, we shall assume that the statistician is given a sequence of numbers δ_n, $\delta_n > 0$ that tends to zero as n tends to infinity.

The "local neighborhoods" of a point θ_0 will then be taken to be of the type $\{\eta : |\eta - \theta_0| \leq \delta_n b\}$ where b will be a number, $b < \infty$.

We know that this is too simple a situation, but it already points out the main features of the theory. Later, in Section 6.9, we discuss possible modifications.

Under such conditions we can study the behavior of a pair $(P_{\eta,n}, P_{\theta_0,n})$ by looking at logarithms $\Lambda_n(\eta; \theta_0) = \log(dP_{\eta,n}/d P_{\theta_0,n})$ and try to approximate $\Lambda_n(\theta_0 + \delta_n\tau; \theta_0)$ with $\eta = \theta_0 + \delta_n\tau$ by a sum of two terms: a term $\tau' S_n$, linear in τ and a term $-\frac{1}{2}\tau' K_n \tau$, quadratic in τ.

The theory is constructed assuming that such quadratic approximations are valid as n tends to infinity in local neighborhoods that, with the rescaling by δ_n, take the form $\{\tau : |\tau| \leq b\}$.

Section 6.2 introduces what are called the LAQ families. They are families that admit quadratic approximations (as $n \to \infty$) where both S_n and K_n are random, depending on the observations and θ_0. For such families, if one possesses auxiliary estimates that already take their values in the local neighborhoods of a true θ_0, one can construct what we call centering variables Z_n. This is done in Section 6.3. It is shown in Proposition 3, Section 6.3, that these Z_n, together with estimates of the matrices K_n, yield asymptotically sufficient statistics.

Section 6.4 describes the local behavior of posterior distributions. Here we use "local" prior measures, that is, prior measures that concentrate on the neighborhoods $\{\eta : |\eta - \theta_0| \leq \delta_n b\}$ as before. It is shown that, starting with gaussian priors, the true posterior distributions can be approximated by gaussian ones. This is true under the LAQ conditions. It is perhaps not surprising, but the proof is not trivial.

In Section 6.5 we deal with invariance properties. It is shown that the rescaling by numbers δ_n that tend to zero automatically implies that certain invariance properties hold almost everywhere, the "almost" referring to the Lebesgue measure. For example, estimates T_n such that $\delta_n^{-1}(T_n - \theta)$ has a limiting distribution for almost every θ must also be such that for almost every θ, $\delta_n^{-1}(T_n - \theta - \delta_n\tau)$ has the same limiting distribution under $\theta + \delta_n\tau$. This is often assumed as a "regularity property" of estimates.

The invariance properties have further consequences: If families $\{P_{\theta,n}; \theta \in \Theta\}$ satisfy LAQ at all θ, they must also satisfy the more restrictive LAMN conditions at almost all θ.

The LAMN conditions differ from the LAQ ones in that the limiting distribution of the matrices K_n, if it exists, does not depend on the local parameter τ. That is, $\mathcal{L}[K_n|\theta + \delta_n\tau]$ has a limit independent of τ. This implies a special status for the LAMN conditions and the stronger LAN conditions. They are described and studied in Section 6.6, where we prove two major results: a form of the Hájek–Le Cam asymptotic minimax theorem and a form of Hájek's convolution theorem. Because of its potential applications to nonparametric problems we also state a form of the Moussatat-Millar extension of the convolution theorem to infinite-dimensional parameter spaces.

Section 6.8 gives some indications of the possible uses of our centering variables Z_n for the construction of confidence ellipsoids or similar objectives.

Sections 6.7 and 6.9 are in the nature of discussions about what conditions should be used if one wants to obtain more general results. They are included to make the reader aware of the fact that much of the theory has already been extended to cases that are not restricted to the open subsets of $\Re^k$ or to the use of single numbers δ_n that tend to zero.

This chapter uses only linear-quadratic approximations to the loglikelihood ratios. It is possible to consider cases where it is more satisfactory to use *smooth* approximations by functions other than quadratic. For an example, see Le Cam (Prague paper 1973) [1986] pp 391-398. A construction based on minimum contrasts will not usually lead to asymptotically sufficient estimates unless the contrast function closely mimics the loglikelihood function at least locally. The literature on minimum contrast estimates is extensive and cannot be summarized here. See S. van de Geer [2000] for further information.

6.2 Locally Asymptotically Quadratic Families

For each integer n let $\{P_{\theta,n} : \theta \in \Theta\}$ be a family of probability measures on some space $(\mathcal{X}_n, \mathcal{A}_n)$. Let $\delta_n > 0$ be a number. It will be assumed through this section that the following hold:

(O) The set Θ is a subset of a euclidean space $\Re^k$ and the "true" value of θ is situated in an open subset of $\Re^k$ contained in Θ.

(C) Two sequences $\{P_{\theta+\delta_n t_n, n}\}$, $\{P_{\theta,n}\}$ such that $|t_n|$ stays bounded are contiguous.

The first assumption (O) needs to be modified for many problems, but it simplifies life. The choice of scaling by a number δ_n

is also quite restrictive. One could scale by matrices δ_n instead without much ado. One could also make the scaling depend on θ, using $\delta_n(\theta)$. We shall not do so here for simplicity, but we will discuss the matter briefly later in this chapter.

Another assumption that will be made here is that we are interested in the local behavior of the experiment. That means that the numbers δ_n tend to zero and that we are interested in the family $\mathcal{F}_{n,\theta} = \{P_{\theta+\delta_n t,n}; t \in \Re^k, \theta + \delta_n t \in \Theta\}$ or subfamilies of that where $|t|$ stays bounded. Note that since we assume that θ is interior to Θ, the set $\{\theta + \delta_n t; |t| \leq b\}$ will eventually be contained in Θ. In view of this we shall often omit the statement that $\theta+\delta_n t_n$ must belong to Θ, even though it is always implied.

Let $\Lambda_n(\theta + \delta_n t, \theta)$ be the logarithm of likelihood ratio, as defined in (3.1) of Chapter 3:

$$\Lambda_n(\theta + \delta_n t, \theta) = \log \frac{dP_{\theta+\delta_n t,n}}{dP_{\theta,n}}.$$

Definition 1. *[LAQ]. The family $\mathcal{E}_n = \{P_{\eta,n}; \eta \in \Theta\}$ is called locally asymptotically quadratic (LAQ) at θ if there are random vectors S_n and random matrices K_n such that*

(i) condition (C) is satisfied,

(ii) for any bounded sequence $\{t_n\}$ the differences

$$\Lambda_n(\theta + \delta_n t_n, \theta) - [t_n' S_n - \frac{1}{2} t_n' K_n t_n]$$

tend to zero in $P_{\theta,n}$ probability,

(iii) The matrices K_n are almost surely positive definite. Further, if $\mathcal{L}(K)$ is a cluster point of $\mathcal{L}(K_n|P_{\theta,n})$, then K is almost surely positive definite (invertible).

Of course S_n may depend on the fixed point θ and so may K_n. They should be written $S_n(\theta)$ and $K_n(\theta)$. However, as long as it is understood that θ is fixed we shall not use the more cumbersome notation. The condition that K_n is positive definite is added here for convenience. As we shall see in Lemma 1, K_n is almost automatically positive semidefinite eventually. However, the positive definiteness, which implies in particular that K_n has an inverse, will be a convenience.

Definition 2. *[LAMN]. This is locally asymptotically mixed normal as in Definition 1, but with an added restriction: Let $\mathcal{L}(S, K)$ be a cluster point of the sequence of distributions $\mathcal{L}(S_n, K_n|P_{\theta,n})$. Then, given K, the vector S is normal such that $E[t'S]^2 = t'Kt$.*

We shall see later some different but equivalent definitions.

Definition 3. *[LAN]. Locally asymptotically normal is the same as LAQ or LAMN, but it is assumed that in the cluster points of $\mathcal{L}(K_n|P_{\theta,n})$ the matrices K are nonrandom .*

It is clear that LAN implies LAMN and that, in turn, LAMN implies LAQ.

There are a multitude of examples where the LAN conditions are satisfied. One of the most classical ones will be discussed in Chapter 7. For the time being, let us mention a simple example exhibiting the difference between the three definitions.

Let $\{X_j, j = 1, 2, \ldots\}$ be a sequence of independent normal random variables, each with expectation μ_ν and variance σ_ν^2. (Here, ν does not have any function, except tending to infinity.) Assume the variances are known and small. Let $S_n = \sum_{j \leq n} X_j$. One obtains an instance of LAN by looking at the X_j or the S_n up to a fixed number m chosen in advance so that $m\mu_\nu$ and $m\sigma_\nu^2$ stay bounded, the last quantity also staying away from zero.

One obtains a case satisfying LAMN by selecting m at random, independently of the $\{X_j\}$, subject to similar boundedness conditions (but in probability).

To obtain an example of LAQ that is not LAMN, stop the sequence S_n the first time it gets out of an interval $[-a, +a]$. (Think of the μ_ν and σ_ν as very small compared to a, so that the experiment is essentially one on brownian motion with drift.) As long as the S_n stay in $[-a, +a]$, they have the usual gaussian distribution, but restricted to $[-a, +a]$. At exit time, they take values that are close to either $-a$ or $+a$ and therefore cannot be mixtures of gaussian as required by LAMN. Further examples using stochastic processes may be found in the book by Basawa and Prakasa Rao [1980] or Basawa and Scott [1983] or in Jeganathan's papers [1980, 1983, 1988].

The approximability of the log likelihood assumed in the LAQ condition suggests that the pair (S_n, K_n) may have some asymptotic sufficiency properties, at least for the experiments $\mathcal{F}_{n,\theta} = \{P_{\theta+\delta_n t,n}; t \in \Re^k\}$ where θ stays fixed. Because S_n is really $S_n(\theta)$ and K_n is $K_n(\theta)$ this may not be so interesting. They are not "statistics." We shall prove a stronger property below.

For the time being, let us note some properties that are implied by conditions (i) and (ii) of the LAQ definition (using (O) as needed but *not* (iii)).

First note an additivity property: If the measures $P_{\theta+\delta_n t,n}$; $t \in \Re^k$ were mutually absolutely continuous, one could write

$$\Lambda_n[\theta + \delta_n(s+t); \theta + \delta_n s] + \Lambda_n[\theta + \delta_n s; \theta] = \Lambda_n[\theta + \delta_n(s+t); \theta]$$

almost everywhere $P_{\theta,n}$. The contiguity assumption allows us to discard pieces of $P_{\theta+\delta_n t,n}$ that are singular with respect to $P_{\theta,n}$. Thus the additivity restriction will still hold except in sets whose probability tends to zero as $n \to \infty$.

Note also that one can retrieve an approximation to K_n by taking differences of the function Λ_n. For instance, if $u_1, u_2, \ldots, u_k$ is a basis for $\Re^k$ one can look at the differences

$$\Lambda_n[\theta + \delta_n(u_1 + u_2); \theta] - \Lambda_n[\theta + \delta_n u_1; \theta] - \Lambda_n[\theta + \delta_n u_2; \theta]. \quad (6.1)$$

If (i) and (ii) hold this must be approximately

$$-\frac{1}{2}\{(u_1 + u_2)' K_n (u_1 + u_2) - u_1' K_n u_1 - u_2' K_n u_2\} = -u_1' K_n u_2.$$

The contiguity assumption (C) implies that all variables of the type

$$\Lambda_n[\theta + \delta_n(u_i + u_j); \theta] \quad \text{with } 0 \le i \le k, 0 \le j \le k \text{ and } u_0 = 0$$

must be bounded in probability for any probability of the type $P_{\theta+\delta_n t_n,n}$ with t_n bounded. Thus K_n and consequently S_n are bounded in $P_{\theta,n}$ probability. Thus, *the joint distributions* $\mathcal{L}[S_n, K_n|\theta]$ *form a relatively compact sequence.*

Suppose that one extracts a subsequence, say, $\{n(\nu)\}$, such that $\mathcal{L}[S_{n(\nu)}, K_{n(\nu)}|\theta]$ converge to a limit $\mathcal{L}(S, K)$. Then by (ii) and contiguity (see Chapter 3), one must have

$$E\exp[t'S - \frac{1}{2}t'Kt] = 1$$

for all t. In particular, applying this to t and $(-t)$ one gets

$$E(\cosh t'S)\exp[-\frac{1}{2}t'Kt] \equiv 1.$$

This has an interesting consequence that shows that condition (iii) of LAQ is "nearly" implied by (i) and (ii).

Lemma 1. *Let conditions (O), (i), and (ii) of LAQ be satisfied. Then in all the cluster points $\mathcal{L}(S, K)$ of $\mathcal{L}(S_n, K_n|\theta)$ the random matrices K are almost surely positive semidefinite.*

Proof. Put a norm on $\Re^k$ and use on the matrices the corresponding norm. For instance, $\|K\| = \sup\{\|Kt\|; \|t\| = 1\}$. We shall prove the lemma by contradiction. Suppose, that there is some $\eta > 0$ such that $\inf_t t'Kt < -2\eta$ with probability at least $2\epsilon > 0$. Eliminating cases of probability at most ϵ, one can assume that $\|K\|$ stays bounded.

Thus, using the compactness of the unit ball of $\Re^k$ one can assume that there is a finite set, say, $\{t_j; j = 1, 2, \ldots, m\}$, with $\|t_j\| = 1$ and $t_j'Kt_j < -\eta$ for some j. This implies that there is at least one t_j, say, v, such that the probability that $v'Kv \leq -\eta$ is strictly positive.

Let A be the set where $v'Kv \leq -\eta$. Then

$$EI_A[\cosh t'S]\exp[-\frac{1}{2}t'Kt] \leq E[\cosh t'S]\exp[-\frac{1}{2}t'Kt] = 1$$

for all multiples t of v.

This is impossible unless $EI_A = 0$, concluding the proof of the lemma.

Thus K must be almost surely positive semidefinite. □

One can also say that the experiments $\mathcal{E}_n$ are K_n-controlled at θ in the sense of definition 3, chapter 11, section 4, of Le Cam [1986, p. 221]. Here this means that if t_n is a bounded sequence such that $t_n' K_n t_n$ tends to zero in $P_{\theta,n}$ probability, then $t_n' S_n$ does the same. This can be seen as above taking subsequences $n(\nu)$ such that $\mathcal{L}[S_{n(\nu)}, K_{n(\nu)} | P_{\theta,n}]$ converges.

We shall return later to the relations between the LAQ conditions and the convergence of experiments described in Chapters 3 and 5. For now, we shall describe a method of construction of centering variables Z_n that, together with estimates of the matrices K_n, will have asymptotic sufficiency properties. It is based on the relations between differences of log likelihood described in (6.1) above.

6.3 A Method of Construction of Estimates

For each integer n, consider an experiment $\mathcal{E}_n = \{P_{\theta,n}; \theta \in \Theta\}$ where Θ is a subset of a euclidean space $\Re^k$. It often happens that the statistician has available some auxiliary estimate θ_n^* that indicates in what range θ is likely to be. The question might then be how to use this θ_n^* to get a better summary of the information available in the experiment, as well as a possibly better estimate. Here we describe some properties of a method that yields, under the LAQ conditions, estimates with asymptotic optimality properties. See, Proposition 3 and Section 6.4 on Bayes procedures. The method amounts to the following: Fit a quadratic to the log likelihood around the estimated value θ^* and take as the new estimated value of θ the point that maximizes the fitted quadratic. To ensure that the method works, we shall use two assumptions; the first says that θ_n^* puts you in the right neighborhood.

Assumption 1. *Let θ be the true value. Then, given $\epsilon > 0$, there are numbers $b(\theta, \epsilon)$ and $n(\theta, \epsilon)$ such that if $n \geq n(\theta, \epsilon)$ one has $P_{\theta,n}\{|\theta_n^* - \theta| \geq \delta_n b(\theta, \epsilon)\} \leq \epsilon$.*

Our next assumption is purely technical. It involves subsets of Θ that, for want of a better name, we shall call δ_n-sparse. Subsets $\Omega_n \subset \Theta$ will be called δ_n-sparse if: (a) for any fixed $b > 0$, any ball of radius $b\delta_n$ contains a finite number of elements of Ω_n, a number that remains bounded independently of n, and (b) there is a fixed number c, such that any point in Θ is within distance $c\delta_n$ of some element of Ω_n.

Assumption 2. *(Discretization) One has preselected δ_n-sparse subsets Ω_n. The auxiliary estimate θ_n^* takes its values in Ω_n.*

This may sound strange, but is easy to enforce. If one has any auxiliary estimate, say, θ_n', one recomputes it, keeping $37 + |\log_{10} \delta_n|$ decimals for each coordinate, give or take one or two decimals. Assumption 2 will be used to avoid *uniform* convergence restrictions, as explained later.

Let $u_0 = 0$ and let $u_1, \ldots, u_k$ be a basis of $\Re^k$. Compute log likelihoods of the form

$$\Lambda_n(\theta_n^* + \delta_n(u_i + u_j); \theta_n^*)$$

for $i, j = 0, 1, 2, \ldots, k$. This gives just enough points to fit a linear-quadratic form with a symmetric matrix for the quadratic. That is, one looks for a function of t

$$t'L_n - \frac{1}{2}t'M_n t$$

that has a linear term (in t) and a quadratic term and takes the value $\Lambda_n[\theta_n^* + \delta_n(u_i + u_j); \theta_n^*]$ if t is equal to $u_i + u_j$. This can easily be done as follows. Take two vectors u_i, u_j with i and j different from zero. Then an "inner product" $u_i' M_{n,i,j} u_j$ between u_i and u_j can be obtained from

$$\begin{aligned} M_{n,i,j} = & - \{\Lambda_n[\theta_n^* + \delta_n(u_i + u_j); \theta_n^*] - \Lambda_n[\theta_n^* + \delta_n u_i; \theta_n^*] \\ & - \Lambda_n[\theta_n^* + \delta_n u_j; \theta_n^*]\}. \end{aligned}$$

This gives the matrix of the quadratic term. The linear terms in $t'L_n$ can then be computed from

$$u_j' L_n = \Lambda_n[\theta_n^* + \delta_n u_j, \theta_n^*] + \frac{1}{2} M_{n,j,j}.$$

If the matrix $M_n = \{M_{n,i,j}\}$, $i, j = 1, 2, \ldots, k$ has an inverse, one can write these last relations in the form

$$\delta_n^{-1}(Z_n - \theta_n^*)' M_n u_j = \Lambda_n(\theta_n^* + \delta_n u_j; \theta_n^*) + \frac{1}{2} M_{n,j,j},$$

where we have just written

$$L_n = \delta_n^{-1} M_n (Z_n - \theta_n^*)$$

for the linear component. If one desires, one can write the same relation in the form

$$Z_n = \theta_n{}^* + \delta_n M_n^{-1} L_n \tag{6.2}$$

where Z_n appears as θ_n^* with a correction added.

The Z_n so defined are the centering variables that will be used as estimates of θ. The matrices M_n give some idea of the precision of the estimates.

Note that if the M_n are positive definite, we obtain a linear quadratic expression of the form

$$-\frac{1}{2}\delta_n^{-2}[Z_n - (\theta_n^* + \delta_n t)]' M_n [Z_n - (\theta_n^* + \delta_n t)]$$

$$+\frac{1}{2}\delta_n^{-2}[Z_n - \theta_n^*]' M_n [Z_n - \theta_n^*] = t' M_n \delta_n^{-1}(Z_n - \theta_n^*) - \frac{1}{2} t' M_n t$$

as an approximation of $\Lambda(\theta_n^* + \delta_n t; \theta_n^*)$. This shows that the value of the corrected estimate that maximizes the fitted quadratic is Z_n itself. Thus, the method could also be described as a modification of the maximum likelihood method where one maximizes a smooth local approximation to the log likelihood function.

That does not mean that the method must be specially successful, but we shall see here and in later sections that the Z_n have some remarkable properties, of the kind often said to be reserved to maximum likelihood.

Before we can state a precise result, we have to take some precautions. It has been noted above that logarithms of likelihood ratios satisfy an additivity relation of the type

$$\Lambda_n[\theta + \delta_n(s+t); \theta + \delta_n s] + \Lambda_n[\theta + \delta_n s; \theta] = \Lambda_n[\theta + \delta_n(s+t); \theta]$$

at least almost everywhere $P_{\theta,n}$ if the measures involved at θ, $\theta+\delta_n s$, and $\theta+\delta_n(s+t)$ are mutually absolutely continuous. Here we do not assume mutual absolute continuity, but a contiguity assumption will still make the relation approximately valid, neglecting sets that have a small probability, tending to zero as $n \to \infty$. The method of construction of Z_n is based on the hope that such an additivity relation remains valid if one substitutes for θ an estimate θ_n^* provided that θ_n^* is not constructed specially to look for peculiar features of the likelihood function. Thus the method of fitting a quadratic described earlier should have the following properties.

Suppose that in the selected range (of order δ_n) around θ_n^* the log likelihood does admit a good quadratic approximation. Suppose also that θ_n^* does not look at special features of the log likelihood. Then it should not matter much what values θ_n^* takes. The quadratic to be used will have a maximum located at a point that does not depend very much on the value of θ_n^*. That, of course, is only approximate because the quadratic is only an approximation. But it amounts to saying that functions of t such as $\Lambda_n[\theta_1+\delta_n t; \theta_1]$ and $\Lambda_n[\theta_2+\delta_n t; \theta_2]$ will reach their maximum, if unique, at points t_1 and t_2 such that $\theta_1+\delta_n t_1 = \theta_2+\delta_n t_2$. So the same must be true, approximately, for quadratic approximations if they are good enough.

Unfortunately, there is no reason why the auxiliary estimate θ_n^* would not search for peculiarities in the log likelihood. One often-used estimate is the maximum likelihood. It does search for a peculiarity of the likelihood function, namely, its maximum value. This is not serious in very smooth situations, but think of taking n observations from a density $f(x,\theta) = ce^{-|x-\theta|^2}/|x-\theta|$ where x and θ are both in $\Re^4$. A maximum, equal to $+\infty$, is attained at each x_j in the sample. This would make a maximum likelihood estimatie (m.l.e.) argument and construction very unhealthy.

It turns out that, to prove limit theorems, one can prevent such misbehavior by a simple trick: It is sufficient to discretize θ_n^* as provided by Assumption 2. This, of course, is not the only possibility; some other techniques are discussed in Le Cam

[1960, appendix I].

When such conditions are satisfied one can prove results as follows.

Proposition 1. *Let $\{P_{\eta,n}; \eta \in \Theta\}$ be a family of probability measures on σ-fields $\mathcal{A}_n$. Let θ_n^* be auxiliary estimates satisfying Assumptions 1 and 2. Let (Z_n, M_n) be the centerings and matrices constructed as described. Then at every θ that satisfies the LAQ conditions (with vectors and matrices S_n and K_n)*

(i) the differences $\delta_n^{-1}(Z_n - \theta) - K_n^{-1} S_n$ and $M_n - K_n$ tend to zero in $P_{\theta,n}$ probability, and

(ii) $\delta_n^{-1}(Z_n - \theta)$ remains bounded in $P_{\theta,n}$ probability.

Remark 1. The δ_n of the construction is supposed to be the same as that of the LAQ condition.

Remark 2. Note that the proposition does not say anything about points θ where the LAQ condition is not satisfied.

Remark 3. The S_n and K_n of the LAQ assumption depend on point θ. This point is unknown in a statistical situation. By contrast, the Z_n and M_n are computable statistics, not involving unknown parameters.

Proof. The quadratic approximation in the LAQ property can be put in the form

$$\Lambda_n(\theta + \delta_n t_n; \theta) \sim -\frac{1}{2}\{(T_n - t_n)' K_n (T_n - t_n) - T_n' K_n T_n\}$$

with $T_n = K_n^{-1} S_n$. If $\{s_n\}$ is another bounded sequence, the additivity relation for log likelihoods and the contiguity included in LAQ show that

$$\begin{aligned} r(n, s_n) &= \Lambda_n[\theta + \delta_n s_n + \delta_n t_n; \theta + \delta_n s_n] \\ &\quad + \frac{1}{2}\{(T_n - s_n - t_n)' K_n (T_n - s_n - t_n) \\ &\quad - (T_n - s_n)' K_n (T_n - s_n)\} \end{aligned}$$

will also tend to zero in probability.

It is tempting to apply this relation, replacing the nonrandom s_n by a random s_n^* such that $\theta_n^* = \theta + \delta_n s_n^*$. However, this may fail. It is possible for a sequence of random functions, say, φ_n, to tend to zero in such a way that $\varphi_n(t_n) \to 0$ in probability for any *nonrandom* sequence $\{t_n\}$ such that $|t_n| \leq b$ but also in such a way that $\sup\{|\varphi_n(t)|;\ |t| \leq b\}$ does not tend to zero in probability. In fact this is common behavior.

Nevertheless, by Assumption 1, given an $\epsilon > 0$, one can find a $b < \infty$ such that $P_{\theta,n}\{|\theta_n^* - \theta| \geq \delta_n b\} \leq \epsilon$ for large n. In the range $\{\eta : |\eta - \theta| \leq \delta_n b\}$ the number of different values taken by the estimate θ_n^* is bounded independently of n, and the possible values are points of Ω_n, which was prescribed in advance. Thus we are looking only at a bounded number of possible random variables, and one may conclude that

$$\Lambda_n[\theta + \delta_n s_n^* + \delta_n t_n; \theta + \delta_n s_n^*]$$

$$+\frac{1}{2}\{[T_n - s_n^* - t_n]' K_n [T_n - s_n^* - t_n] - [T_n - s_n^*]' K_n [T_n - s_n^*]\}$$

will also tend to zero in probability for all choices of t_n of the form $t_n = u_i + u_j$, $i, j = 0, 1, \ldots, k$. Compare this to the equations that determine Z_n and M_n. They can be put in the form

$$\Lambda_n[\theta + \delta_n s_n^* + \delta_n t_n; \theta + \delta_n s_n^*]$$

$$= -\frac{1}{2}\{[\delta_n^{-1}(Z_n - \theta) - s_n^* - t_n]' M_n [\delta_n^{-1}(Z_n - \theta) - s_n^* - t_n]$$
$$-[\delta_n^{-1}(Z_n - \theta) - s_n^*]' M_n [\delta_n^{-1}(Z_n - \theta) - s_n^*]\}$$

with $t_n = u_i + u_j$ as usual. Taking second differences and comparing terms, we see that $M_n - K_n$ must tend to zero in probability. Since K_n is positive definite, with positive-definite cluster points, the same will be true eventually of M_n except in cases whose probability tends to zero. Looking then at the linear terms, one sees that $M_n \delta_n^{-1}(Z_n - \theta) - K_n T_n$ will tend to zero in probability. By construction, $K_n T_n = S_n$.

So far we have not used condition (iii) of the LAQ definition except incidentally. Now let us consider $T_n = K_n^{-1} S_n$. We claim that it is also bounded in $P_{\theta,n}$ probability. Indeed if a subsequence of $\mathcal{L}(S_n, K_n|P_{\theta,n})$ tends to a limit $\mathcal{L}(S, K)$ then K is almost surely positive definite and hence invertible. Thus K_n^{-1} is bounded in probability, and the same is true of $T_n = K_n^{-1}S_n$. In addition since $M_n - K_n \to 0$ in probability, so does $K_n^{-1}M_n - I$. Thus $\delta_n^{-1}(Z_n - \theta) - T_n$ is also bounded in probability, and so is $[\delta_n^{-1}(Z_n - \theta)]'M_n[\delta_n^{-1}(Z_n - \theta)]$.

This concludes the proof of the proposition. □

In fact, it will be convenient to have a technical reinforcement of Proposition 1 as follows.

Proposition 2. *Let θ be a point at which LAQ holds. Let b and $\epsilon > 0$ be given positive numbers. Then there exist numbers $c = c(b, \epsilon)$ and $N = N(b, \epsilon)$ such that, if $n \geq N$ and $|\eta - \theta| \leq \delta_n b$, then*

$$P_{\eta,n}\{[|Z_n - \theta| > \delta_n c] \cup [\|M_n\| > c]\} < \epsilon. \tag{6.3}$$

Furthermore, if I_n is the indicator of the set $[|Z_n - \theta| \leq \delta_n c] \cap [\|M_n\| \leq c]$ and if $Q_{\eta,n}$ has density

$$\exp\left[-\frac{1}{2}\delta_n^{-2}\{[Z_n - \eta]'M_n[Z_n - \eta] - [Z_n - \theta]'M_n[Z_n - \theta]\}\right] \tag{6.4}$$

with respect to $P_{\theta,n}$, then $\int I_n|dQ_{\eta,n} - dP_{\eta,n}|$ tends to zero uniformly for $|\eta - \theta| \leq b\delta_n$ as n tends to infinity.

Proof. The existence of the numbers c and N follows from Proposition 1 for the measure $P_{\theta,n}$ itself. The result as stated for the sets $\{|\eta - \theta| \leq \delta_n b\}$ follows by contiguity. See, for example, property (b) in Theorem 1, Section 3.1. For the second statement, we already know by Proposition 1 that

$$I_n[\frac{dQ_{\eta,n}}{dP_{\theta,n}} - \frac{dP_{\eta,n}}{dP_{\theta,n}}]$$

tends to zero in probability. However, if multiplied by I_n, the first likelihood ratio stays bounded. The second behaves according to the contiguity requirement. That means that, given an

$\epsilon > 0$, there is a b such that the part of $dP_{\eta,n}/dP_{\theta,n}$ above b contributes at most ϵ to $P_{\eta,n}$. This is enough to imply the desired result.□

Note 1. The measures $Q_{\eta,n}$ may happen to be infinite, but their restrictions when multiplied by I_n are finite.

Proposition 3. *The pair (Z_n, M_n) is asymptotically sufficient in the following sense.*

There are other families of probability measures $\{R_{\eta,n}; \eta \in \Theta\}$ defined on the σ-field $\mathcal{A}_n$ such that

(i) for $\{R_{\eta,n}; \eta \in \Theta\}$ the pair (Z_n, M_n) is sufficient (exactly) and

(ii) for every θ that satisfies LAQ the difference

$$\sup_{|\eta-\theta|\le\delta_n b} \|R_{\eta,n} - P_{\eta,n}\|$$

tends to zero for each fixed b.

Remark 4. The sufficiency is for the entire family $\{R_{\eta,n}; \eta \in \Theta\}$. This is much stronger than sufficiency for a restricted family $\{R_{\eta,n}; |\eta - \theta| \le \delta_n b\}$. An example shows what could happen.

Example 1. Take for Θ the union of the open intervals $(2k + (1/4), 2k + (3/4))$, $k = 0, 1, 2, \ldots$. For $\theta \in \Theta$, let $P_{\theta,n}$ be the joint distribution of i.i.d. normal variables $X_j; j = 1, 2, \ldots, n$, with variance unity and expectation θ. Let F_n be the fractional part of the average $\bar{X}_n = \sum_j^n X_j/n$. Here, F_n is asymptotically sufficient on each interval $(2k + (1/4), 2k + (3/4))$ but it does not satisfy the requirements of Proposition 3. It is unable to sort out in which interval θ lies.

The proof that follows amounts to creating a conditional expectation operator called H_n that maps $\mathcal{A}_n$ measurable functions into $\mathcal{B}_n$ measurable functions, where $\mathcal{B}_n$ is the σ-field generated by (Z_n, K_n). One defines the $R_{\eta,n}$ so that, for bounded $\mathcal{A}_n$-measurable functions u, $\int u dR_{\theta,n} = \int (H_n u) dR_{\theta,n}$. The maps H_n are constructed piecewise in such a way that "near" a θ

satisfying LAQ they are very close to conditional expectations given $\mathcal{B}_n$ for $P_{\theta,n}$.

Proof. The proof will be carried out under the *supplementary assumption* that the $P_{\theta,n}$ are *mutually absolutely continuous.* This is just to avoid sundry technical difficulties. The assumption will be lifted at the end of the proof.

Pave the space $\Re^k$ with cubes whose sides are δ_n long, removing appropriate faces to obtain a partition. Let $C_{\nu,n}$, $\nu = 1, 2, \ldots$ be these cubes. If an intersection $\Theta \cap C_{\nu,n}$ is not empty, select a point $\theta_{\nu,n}$ in it. If $\Theta \cap C_{\nu,n}$ is empty, select a $\theta_{\nu,n} \in \Theta$ "close to" $C_{\nu,n}$. For each $\theta_{\nu,n}$ so selected, let $H_{\nu,n}$ be the conditional expectation given $\mathcal{B}_n$ for $P_{\theta_{\nu,n},n}$. Here conditional expectations will be defined as positive linear projections of the space $\mathcal{M}_{0,n}$ of equivalence classes of bounded $\mathcal{A}_n$-measurable functions onto the space $\mathcal{M}_{1,n}$ of equivalence classes of $\mathcal{B}_n$-measurable functions. The domains of definition will be expanded as usual for unbounded functions. Recall that if H is any positive linear projection of $\mathcal{M}_0$ onto $\mathcal{M}_1$, it automatically satisfies the commutation relation

$$H(uv) = uHv \tag{6.5}$$

for every $v \in \mathcal{M}_0$ and every $u \in \mathcal{M}_1$.

For a proof see Le Cam [1986, p. 59] or the more general arguments of T. Ando [1966].

Now, let $I_{\nu,n}$ be the indicator of the set $\{Z_n \in C_{\nu,n}\}$ and let $H_n = \sum_\nu H_{\nu,n} I_{\nu,n}$. This is a positive linear projection of $\mathcal{M}_{0,n}$ onto $\mathcal{M}_{1,n}$. Define measures $R_{\theta,n}$ by $\int v dR_{\theta,n} = \int (H_n v) dP_{\theta,n}$ for $v \in \mathcal{M}_0$.

For the measures $R_{\theta,n}$, the projection H_n serves as a common conditional expectation so that $\mathcal{B}_n$ is clearly sufficient. Let us show that in suitable vicinities of a θ for which LAQ holds, the second statement of the proposition is valid, or, equivalently, that H_n provides an approximation to the conditional expectations for $P_{\eta,n}$ if $|\eta - \theta| < \delta_n b$. Let V be this set of η.

Letting W_n be the set $W_n = \{x : |x - \theta| \leq \delta_n b'\}$ in $\Re^k$, let D be the set of indices ν such that $W_n \cap C_{\nu,n}$ is not empty. This is a finite set. It will be sufficient to show that, on $W_n' = \bigcup C_{\nu,n}; \nu \in D$, our H_n can serve as approximate conditional expectation for all $\eta \in V$. Note that θ was assumed to be interior to Θ, so that eventually all the $C_{\nu,n}$ with $\nu \in D$ are contained in Θ. Hence for all of them, the $\theta_{\nu,n}$ chosen to define $H_{\nu,n}$ is within distance at most $\delta_n(b' + \sqrt{k})$ of θ. For all ξ and η in that range, one can use the fact that $\log(dP_{\eta,n}/dP_{\xi,n})$ is approximated by

$$-\frac{1}{2}\delta_n{}^{-2}\{(Z_n - \eta)'M_n(Z_n - \eta) - (Z_n - \xi)'M_n(Z_n - \xi)\}.$$

On this set and for $|\eta - \theta| \leq \delta_n b$, define measures $Q_{\eta,n}$ as in Proposition 2. For these, all the conditional expectations given $\mathcal{B}_n$ are the same, independent of η. In addition, the differences $\|I_n[Q_{\eta,n} - P_{\eta,n}]\|$ tend to zero. Thus, letting $E_{\theta,n}$ be the conditional expectation given $\mathcal{B}_n$ for $P_{\theta,n}$, one sees that for all $\nu \in D$, the differences $I_n[H_{\nu,n} - E_{\theta,n}]$ tend to zero.

This would complete the proof of the proposition, since ϵ can be made arbitrarily small. However, we still need to remove the supplementary assumption of equivalence of the $P_{\theta,n}; \theta \in \Theta$. This can be done in many ways. One of them is to pave $\Re^k$ with cubes of side δ_n^2 and replace all $P_{\theta,n}$ for θ in such a cube by just one of them. This gives a countable family $\{P_{\xi,n}\}$ dominated by some probability measure μ_n. One then replaces each $P_{\xi,n}$ by $(1 - \delta_n^2)P_{\xi,n} + \delta_n^2\mu_n$. These substitutions will not affect things materially around a θ for which LAQ holds. □

Remark 5. The fact that the approximation property stated in part 2 of Proposition 3 holds only on shrinking neighborhoods of a point makes it relatively weak. However, suppose that the LAQ conditions and Assumptions 1 and 2 hold *uniformly* on a subset S of Θ. Then the L_1-norms $\|P_{\theta,n} - R_{\theta,n}\|$ will tend to zero *uniformly* for $\theta \in S$. This can be deduced from the above proof by noting that the argument does not depend on keeping θ fixed.

Remark 6. In the present situation we have two σ-fields $\mathcal{A}_n$ and $\mathcal{B}_n$ on the same space, with $\mathcal{B}_n \subset \mathcal{A}_n$. This was a situation mentioned in Chapter 2, where one has two "natural" possible definitions of the "loss" sustained in passing from $\mathcal{A}_n$ to $\mathcal{B}_n$. One is the "deficiency" used in Chapter 2 and throughout. Another one, called "insufficiency" in Le Cam [1986], is defined as follows. Let S be a subset of Θ. For the family $\{P_\theta; \theta \in S\}$, the insufficiency of $\mathcal{B}_n$ with respect to $\mathcal{A}_n$ on S is the infimum of the numbers ϵ for which one could find other families of probability measures, say, $\{R_\theta; \theta \in S\}$ on $\mathcal{A}_n$, such that (i) the σ-field $\mathcal{B}_n$ is sufficient for $\{R_\theta; \theta \in S\}$ and (ii)

$$\sup_{\theta \in S} \frac{1}{2} \|R_\theta - P_\theta\| \leq \epsilon.$$

This proof consists of showing that, in shrinking neighborhoods, the insufficiency of $\mathcal{B}_n$ tends to zero and the approximating families $R_{\theta,n}$ are well pieced together. This piecing together could not be carried out in the locally gaussian example that follows the statement of Proposition 3.

6.4 Some Local Bayes Properties

In this section, we shall consider the same structures as in the previous one with experiments $\{P_{\theta,n}; \theta \in \Theta\}$ and pairs (Z_n, M_n). Assumptions 1 and 2 of the previous section are in force. We shall work around a point θ at which the LAQ conditions are duly satisfied. Our main object is to study what happens to *posterior* distributions for *priors* that concentrate on sets of the type $\{|\eta - \theta| \leq \delta_n b\}$ with δ_n the same number that occurs in the LAQ assumptions, in the sense that, given $\epsilon > 0$, there is a b for which the set $\{\eta, |\eta - \theta| > \delta_n b\}$ has prior measure at most ϵ.

This does not mean that more general *priors* should not be considered. In fact, Assumption 1 by itself already implies that for arbitrary *priors*, the corresponding *posteriors* pretty much concentrate on the small vicinities shrinking at the δ_n rate and therefore will often be amenable to arguments quite similar to the ones described below. See Le Cam [1986, p. 335] for precise statements.

Here, the restricted priors will be sufficient to obtain information on the asymptotic behavior of estimates that are "best" in some sense. See the local minimax theorems of Section 6.6 below.

In fact, we shall start with prior measures μ_n that are gaussian with densities, with respect to the Lebesgue measure

$$h_n(\eta) = \frac{\delta_n^{-1}[\det \Gamma_n]^{\frac{1}{2}}}{(2\pi)^{\frac{k}{2}}} \exp\{-\frac{\delta_n^{-2}}{2}(\eta - \theta)'\Gamma_n(\eta - \theta)\} \qquad (6.6)$$

for some positive definite nonrandom matrices Γ_n. It will be seen that the posterior distributions are also approximately gaussian. This has various consequences for the asymptotic behavior of the centerings Z_n.

The formula given for h_n assumes implicitly that Θ is the entire euclidean space $\Re^k$. We shall carry out the computations under that assumption. Since θ satisfying LAQ is an interior point of $\Re^k$, it is not difficult to modify them for such a case.

The use of prior measures implies possibilities of integration. *Thus, even if not said explicitly, it will be assumed that maps such as* $\eta \rightsquigarrow P_{\eta,n}(A)$ *are measurable.*

Proposition 4. *Let* $\mathcal{E}_n = \{P_{\eta.n}; \eta \in \Theta\}$ *be as described and LAQ at* θ. *Take gaussian priors* μ_n *with densities* h_n *as in (6.6). Let* $G_{x,n}$ *be the gaussian measure centered at*

$$\theta + (M_n + \Gamma_n)^{-1}M_n(Z_n - \theta)$$

with inverse covariance matrix $(M_n + \Gamma_n)$. *Let* $F_{x,n}$ *be the posterior distribution of* η *for the prior measure* μ_n *in the experiment* $\mathcal{E}_n$.

Then if $\|\Gamma_n\|$ *and* $\|{\Gamma_n}^{-1}\|$ *stay bounded, the* L_1*-norms* $\|F_{x,n} - G_{x,n}\|$ *tend to zero. The result remains true without the condition of boundedness of* $\|{\Gamma_n}^{-1}\|$ *if* $\|\Gamma_n\|$ *tends to zero sufficiently slowly.*

Before starting the proof, let us state a little general lemma that will be helpful. On the product $\mathcal{X} \times \Theta$ of two measurable spaces, consider two finite positive measures $S_i; i = 1, 2$. Under mild conditions, for instance, the condition that Θ be a Borel subset of a Polish space with its σ-field of Borel sets, they can be "disintegrated" in the form $S_i(dx, d\theta) = S'_i(dx)F_{x,i}(d\theta)$ where S'_i is a marginal on $\mathcal{X}$ and the $F_{x,i}$ are probability measures on Θ. The aim of the following lemma is to provide bounds on the differences $F_{x,1} - F_{x,2}$ of conditional distributions in terms of bounds on the differences $S_1 - S_2$ of the joint measures.

Lemma 2. *The L_1-norms satisfy the inequality*

$$\int \|F_{x,1} - F_{x,2}\|(S'_1 + S'_2)(dx) \leq 4\|S_1 - S_2\|. \tag{6.7}$$

Proof. Let $S = S_1 + S_2$. It can be disintegrated as $S(dx, d\theta) = S'(dx)F_x(d\theta)$. One can also take Radon-Nikodym derivatives and write $S_i(dx, d\theta) = \phi_i(x, \theta)S(dx, d\theta)$. This gives $S'_i(dx) = \psi_i(x)S'(dx)$ with $\psi_i(x) = \int \phi_i(x, \theta)F_x(d\theta)$. Thus, one can write

$$\psi_i(x)F_{x,i}(d\theta) = \phi_i(x, \theta)F_x(d\theta).$$

Now use the following relation, in which we have suppressed the arguments x and $d\theta$,

$$\psi_1 F_1 - \psi_2 F_2 = \frac{1}{2}(\psi_1 + \psi_2)(F_1 - F_2) + \frac{1}{2}(\psi_1 - \psi_2)(F_1 + F_2).$$

Since we have taken $S = S_1 + S_2$ as the dominating measure, we have $\phi_1 + \phi_2 = \psi_1 + \psi_2 = 1$. Taking this into account, the previous identity reads

$$\frac{1}{2}(F_1 - F_2) = [\psi_1 F_1 - \psi_2 F_2] + [\psi_2 - \psi_1]\frac{1}{2}(F_1 + F_2).$$

Now take absolute values and integrate out θ getting

$$\frac{1}{2}\|F_1 - F_2\| \leq \int |\psi_1 F_1 - \psi_2 F_2| + |\psi_2 - \psi_1|.$$

Multiplying this by S' and integrating yields

$$\frac{1}{2}\int \|F_{x,1} - F_{x,2}\|S'(dx) \leq \|S_1 - S_2\| + \|S'_1 - S'_2\|.$$

The desired inequality follows. □

With this we can proceed to the proof of the proposition.

Proof of Proposition 4. It is convenient to change the notation temporarily to avoid the δ_n and some other nuisances. Thus, we shall reparametrize using instead of η the parameter $\tau = \delta_n{}^{-1}(\eta - \theta)$. Similarly, let $z_n = \delta_n{}^{-1}(Z_n - \theta)$. The prior for τ is then the gaussian γ_n centered at 0 with inverse covariance matrix Γ_n. The measures $P_{\theta+\delta_n\tau,n}$ will be called $W_{\tau,n}$.

Proceeding as in Proposition 3, Section 6.3, select a number b so that $\gamma_n[|\tau| \geq b] < \epsilon$. Then there are numbers c and N with the following properties. Let B_n be the set where $|z_n| \leq c$, $\|M_n\| \leq c$, and M_n is positive definite. Then for $n \geq N$ the complement of B_n has probability at most ϵ for all $W_{\tau,n}$ with $|\tau| < b$.

Let then $V_{\tau,n}$ be $W_{\tau,n}$ restricted to B_n, that is, for $A \in \mathcal{A}_n$, one has $V_{\tau,n}(A) = W_{\tau,n}(A \cap B_n)$. Further, introduce other measures, say, $U_{\tau,n}$, whose density with respect to $V_{0,n}$ is

$$\exp\{-\frac{1}{2}[(z_n - \tau)'M_n(z_n - \tau) - z_n'M_n z_n]\}.$$

Note that this expression stays bounded on the set B_n. The LAQ conditions imply, as in Section 6.3, that $\|V_{\tau,n} - U_{\tau,n}\|$ tends to zero as n tends to infinity uniformly for τ bounded. Introduce the joint measures $S_{n,i}$ by

$$S_{n,1}(dx, d\tau) = V_{\tau,n}(dx)\gamma_n(d\tau), \quad S_{n,2}(dx, d\tau) = U_{\tau,n}(dx)\gamma_n(d\tau).$$

It follows that $\|S_{n,1} - S_{n,2}\|$ tends to zero.

The conditional distributions of τ for $S_{n,2}$ are easily seen to be gaussian. In fact, they are the gaussian measures $G'_{x,n}$ obtained from the $G_{x,n}$ of the proposition by passing from the variable η to τ. A similar change of variables gives conditional distributions $F'_{x,n}$ for $S_{n,1}$. At least this applies on the set B_n; outside of that set, what can happen is not specified. One concludes by application of Lemma 2 that $\int \|F'_{x,n} - G'_{x,n}\|S'_{n,1}(dx)$ tends to zero.

Here $S'_{n,1}$ is the marginal of $S_{n,1}$, that is, $\int V_{\tau,n}\gamma_n(d\tau)$. In the original notation, it is the restriction to B_n of $\bar{P}_n = \int P_{\eta,n}\mu_n(d\eta)$. This latter $\bar{P}_n$ and the measures $W_{0,n} = P_{\theta,n}$ are easily seen to be contiguous. Thus one can also say that

$$\int I_{B_n} \|F_{x,n} - G_{x,n}\| P_{\theta,n}(dx) \to 0.$$

If in this last integral one includes the complement of B_n, this would yield

$$\lim_n \int \|F_{x,n} - G_{x,n}\| \leq \lim_n P_{\theta,n}(B_n^c) \leq \epsilon.$$

Since ϵ is arbitrarily small, this concludes the proof, at least when $\|{\Gamma_n}^{-1}\|$ stays bounded.

To obtain the result when Γ_n tends to zero slowly, one can use a standard diagonal procedure. Hence the proposition can be considered proved. □

Remark 7. The preceding proof is not very satisfactory since it does not provide explicit bounds for each given n. This is particularly true for the numbers called ϵ_n and the use of a diagonal argument. It should be possible to get better results in view of the derivations to be found in Le Cam [1986, p. 325–336].

Remark 8. Proposition 4 has several consequences. Suppose we wish to estimate η with a loss function of the type $L_n(t_n - \eta)$ where L_n stays bounded, say, by C, and where the sets $\{\xi : L_n(\xi) \leq \alpha\}$ are convex symmetric around zero. Suppose that $G_{x,n}$ were the true posterior. Then a lemma of Anderson [1955] (see below and Section 6.7) says that for the gaussian posterior, the minimum posterior risk is achieved for

$$t = \theta + (M_n + \Gamma_n)^{-1} M_n (Z_n - \theta).$$

Note that this may be rewritten in the form

$$t = [I - (M_n + \Gamma_n)^{-1} M_n]\theta + (M_n + \Gamma_n)^{-1} M_n Z_n,$$

or, when M_n is nonsingular,

$$t = [I - (I + \Gamma_n M_n^{-1})^{-1}]\theta + [I + \Gamma_n M_n^{-1}]^{-1} Z_n.$$

Thus if Γ_n is small enough, t is close to Z_n.

The minimum posterior risk for the gaussian distributions $G_{x,n}$ is $\int L_n(t-\eta)G_{x,n}(d\eta)$. Since

$$\left|\int L_n(s-\eta)G_{x,n}(d\eta) - \int L_n(s-\eta)F_{x,n}(d\eta)\right| \leq C\|G_{x,n} - F_{x,n}\|$$

for all s, our t will also almost minimize the true posterior risk. If in addition L_n has some reasonable continuity property, the centering Z_n will also achieve a posterior risk that is close to the minimum possible. Integrating x out, one concludes that Z_n almost gives the minimum Bayes risk. We shall give some additional related results in the next section.

Anderson's lemma will be used again in Section 6.7. What it says is the following.

Anderson's Lemma. *Let f be a probability density on $\Re^k$. Let g be a nonnegative function. Assume that for all α the sets $\{x : f(x) \geq \alpha\}$ and $\{x : g(x) \leq \alpha\}$ are convex sets symmetric around zero. Then, λ being the Lebesgue measure,*

$$\int g(x)f(x+y)\lambda(dx) \geq \int g(x)f(x)\lambda(dx) \quad \text{for all} \quad y \in \Re^k.$$

Remark 9. The reader should particularly note that Proposition 4 refers to prior distributions that change with n and concentrate around θ in neighborhoods of the type $\{\eta : |\eta - \theta| \leq \delta_n b\}$. The fact that they are gaussian is not essential to the arguments; it just makes computations easy and more explicit. On the contrary, Proposition 4 does not say anything about what happens for priors that do not concentrate around θ, such as fixed priors independent of n. We shall see later, in Chapter 8, that in such cases we can still obtain results that resemble Proposition 4. This was noted by Laplace around 1810 but was later given the name Bernstein–von Mises phenomenon. One can show that the mere existence of auxiliary estimates θ_n^* satisfying Assumption 1 of Section 6.3 already implies a form of concentration of posterior distributions around the true value θ; see Le Cam [1986, chapter 12, section 3, proposition 3]. However, the situation is rather complex. We shall only treat special cases later, in Chapter 8.

Remark 10. Even though Proposition 4 is limited as explained earlier, it is not entirely trivial. The reader may wish to consider a particular example of densities in $\Re^3$ of the form

$$f(x, \theta) = c\frac{\exp\{-|x-\theta|^2\}}{|x-\theta|^{1/2}}.$$

One takes n independent identically distributed observations $x_1, x_2, \ldots, x_n$. The posterior densities will have infinite peaks at each x_j. Yet Proposition 4 applies, for a rate $\delta_n = n^{-1/2}$. This is further discussed in Chapters 7 and 8.

6.5 Invariance and Regularity

The technique of approximating likelihood ratios around a point θ in neighborhoods that shrink at a certain rate δ_n, $\delta_n > 0$, $\delta_n \to 0$ is very common in statistics. At first sight, it may seem devoid of any great significance, but it has some invariance implications that are quite important. Here we shall deal only with sequences of *numbers* $\delta_n > 0$ that tend to zero and do not depend on θ. This is only for simplicity. One can replace the numbers δ_n by matrices $\delta_n(\theta)$ depending on θ at the cost of some complications, mostly in notation. What counts is that the neighborhoods shrink as n increases.

To describe the situation, take experiments $\mathcal{E}_n = \{P_{\theta,n} : \theta \in \Theta\}$, $\Theta = \Re^k$, and numbers δ_n. One can form an atlas of the system as follows: For each pair (θ, ξ), let $F_n(\theta, \xi)$ be the measure $P_{\theta+\delta_n\xi,n}$. If one fixes θ, ξ can range over $\Re^k$ and the family $\mathcal{F}_{n,\theta} = \{F_n(\theta,\xi); \xi \in \Re^k\}$ is an experiment "locally" attached to θ.

Now, since Θ is taken equal to $\Re^k$, one can apply shift transformations to θ or to ξ, as follows. We shall denote $S_1(\alpha)$ the shift by the amount α operating on the first argument, θ. The shift by α operating on the second argument, ξ, will be $S_2(\alpha)$. Thus

$$F_n[S_1(\alpha)(\theta,\xi)] = F_n(\theta+\alpha, \xi),$$

$$F_n[S_2(\alpha)(\theta,\xi)] = F_n(\theta,\xi+\alpha).$$

Note that, by construction,

$$F_n[S_1(\delta_n\alpha)(\theta,\xi)] = F_n[S_2(\alpha)(\theta,\xi)]$$

since both are the measure $P_{\theta+\delta_n(\alpha+\xi),n}$. The limiting invariance properties come from that equality: It can be written symbolically as

$$\mathcal{F}_n[S_1(\delta_n\alpha)] = \mathcal{F}_n[S_2(\alpha)],$$

for the family $\mathcal{F}_n = \{F_n(\theta,\xi); \theta,\xi \in \Re^k\}$. On the left side, the shift operation $S_1(\delta_n\alpha)$ tends to the identity $S_1(0)$, whereas on the right side, δ_n does not occur. Thus one could expect that if one can take limits $\mathcal{F}_n \to \mathcal{F}$ one could write $\mathcal{F} = \mathcal{F}S_2(\alpha)$. This is a heuristic argument that will be made precise later.

The same thing happens to distributions of statistics or estimates T_n on $\mathcal{E}_n$. Because of the applications to the LAQ case, we shall let $H_n(\theta,\xi)$ be the distribution of the pair $(\delta_n^{-1}(T_n-\theta), K_n)$ for the measure $P_{\theta+\delta_n\xi,n}$. Shifting the entry ξ by the amount α gives us the distribution of $(\delta_n^{-1}(T_n-\theta), K_n)$ at $\theta+\delta_n(\alpha+\xi)$.

Shifting θ by $\delta_n\alpha$ gives us the distribution of $(\delta_n^{-1}(T_n-\theta-\delta_n\alpha), K_n)$ for the measures $P_{\eta,n}$ at $\eta = \theta+\delta_n(\alpha+\xi)$. Note that α occurs in the random elements only by the combination $\delta_n^{-1}\delta_n\alpha = \alpha$. Note also that K_n is not shifted. The measures H_n can also be shifted as usual: If H_n is the distribution of a variable X then the symbol $S^\alpha H_n$ stands for the distribution of $X+\alpha$.

With this symbolism one can write

$$\mathcal{L}\{(\delta_n^{-1}(T_n-\theta-\delta_n\alpha), K_n)|P_{\theta+\delta_n(\alpha+\xi),n}\}$$

$$= H_n(\theta+\delta_n\alpha,\xi) = S^{-\alpha}H_n(\theta,\alpha+\xi).$$

Passing to the limit as before, one would conclude that the limit H satisfies $H(\theta,\xi) = S^{-\alpha}H(\theta,\alpha+\xi)$. Equivalently the limiting distribution of $(\delta_n^{-1}(T_n-\theta-\delta_n\alpha), K_n)$ under $P_{\theta+\delta_n(\alpha+\xi),n}$ is the same as if α was taken equal to zero.

These passages to the limit seem to involve quite a bit of wishful thinking. However, one can certainly justify them under a variety of conditions. One of them is so-called continuous convergence: The functions $(\theta, \xi) \rightsquigarrow H_n(\theta, \xi)$ converge continuously to a limit $(\theta, \xi) \rightsquigarrow H(\theta, \xi)$ if whenever $\theta_n \to \theta_0$, then $H_n(\theta_n, \xi)$ tends to $H(\theta_0, \xi)$. There are weaker conditions that will do, for instance, one can take only sequences θ_n tending to θ_0 at the rate δ_n.

Here we shall describe a way of taking limits by averaging the $H_n(\theta, \xi)$ taking $\int H_n(\theta, \xi)\mu(d\theta)$ for probability measures μ that are absolutely continuous with respect to the Lebesgue measure of $\Re^k$.

This procedure has been used extensively (see Le Cam [1974], Jeganathan [1982, 1983]). Averaging means that the average must be defined. Thus we shall have to assume that the maps $\theta \rightsquigarrow P_{\theta,n}(A)$ are Lebesgue measurable. However we shall restrict ourselves to a particular case to avoid excessive sets of Lebesgue measure zero. It is an assumption that "almost" ensures measurability. It is often satisfied. For instance, it is satisfied under the LAQ conditions. It is as follows.

Definition 4. *The experiments $\mathcal{E}_n = \{P_{\theta,n}; \theta \in \Theta\}$ will be called δ_n-tail continuous at the point θ_0 if the L_1-norm $\|P_{s_n,n} - P_{t_n,n}\|$ tends to zero for all pairs (s_n, t_n) such that*

(i) $\delta_n^{-1}(s_n - \theta_0) + \delta_n^{-1}(t_n - \theta_0)$ stays bounded and

(ii) $\delta_n^{-1}(s_n - t_n) \to 0$.

If this happens at every $\theta_0 \in \Theta$ we shall just say that the $\mathcal{E}_n$ are δ_n-tail continuous.

Note that this property automatically implies an analogous property for our measures H_n: If ξ_n and τ_n are bounded and $\xi_n - \tau_n \to 0$, then

$$\|H_n(\theta_0, \xi_n) - H_n(\theta_0, \tau_n)\| \to 0.$$

To go further we must select a sensible mode of convergence for the measures H_n. They are the distribution of a pair (T_n, K_n) in a euclidean space $\Re^m$ with $m = k + (1/2)k(k+1)$ dimensions. For measures on that space we shall say that (T_n, K_n) tends in distribution to (T, K) if $Eu(T_n, K_n) \to Eu(T, K)$ for all continuous functions u that are bounded and tend to zero as (t, k) tends to infinity in norm.

Applying this to the functions $(\theta, \xi) \rightsquigarrow H_n(\theta, \xi)$, we shall say that the θ-averages of H_n converge to those of H if

$$\int H_n(\theta, \xi)\mu(d\theta) \to \int H(\theta, \xi)\mu(d\theta)$$

in the above sense for measures on $\Re^m$ for all probability measures μ that are dominated by the Lebesgue measure and for all $\xi \in \Re^k$.

It can be shown that this mode of convergence is metrizable for δ_n–tail continuous experiments. It can even be shown that δ_n–tail continuous experiments form relatively compact sequences for that metric. We shall not prove that here. See Le Cam [1986, p. 135, 142]. It is useful to know since it implies that one can extract convergent subsequences, so Proposition 5 applies to such subsequences.

Proposition 5. *Assume that the experiments $\mathcal{E}_n = \{P_{\theta,n}; \theta \in \Re^k\}$ are δ_n–tail continuous and that the $P_{\theta,n}$ are measurable in θ. Assume also that the joint distributions H_n converge to a limit H in the sense of θ-averages as just described. Then there is a set N of Lebesgue measure zero such that if $\theta \notin N$ then*

$$S^\alpha H(\theta, \xi) = H(\theta, \xi + \alpha)$$

for all ξ and α.

Proof. Let us note the equalities

$$\mathcal{L}\{[\delta_n^{-1}(T_n - \theta - \delta_n\alpha), K_n]|P_{\theta+\delta_n(\alpha+\xi)}\}$$
$$= H_n(\theta + \delta_n\alpha, \xi) = S^{-\alpha}H_n(\theta, \alpha + \xi),$$

and

$$\int H_n(\theta + \beta, \xi)\mu(d\theta) = \int H_n(\theta, \xi)(S^\beta\mu)(d\theta).$$

Integrating the first relations we obtain

$$\int H_n(\theta + \delta_n\alpha, \xi)\mu(d\theta) = \int H_n(\theta, \xi)(S^{\delta_n\alpha}\mu)(d\theta)$$

$$= S^{-\alpha} \int H_n(\theta, \alpha + \xi)\mu(d\theta).$$

For the shifted measures, the L_1-norms always satisfy the relation

$$\| \int H_n(\theta, \xi)(S^\beta \mu)(d\theta) - \int H_n(\theta, \xi)\mu(d\theta)\| \le \|S^\beta \mu - \mu\|$$

because the norms $\|H_n(\theta, \xi)\|$ do not exceed unity. Here the shift β is taken equal to $\delta_n \alpha$. It tends to zero. Since μ is dominated by the Lebesgue measure, $\|S^{\delta_n \alpha}\mu - \mu\| \to 0$. Indeed, if μ has density g with respect to the Lebesgue measure λ, one can approximate g by a uniformly continuous function, say, u, so that $\int |g(x) - u(x)|\lambda(dx) = \int |g(x+\beta) - u(x+\beta)|\lambda(dx) < \epsilon$. Because of the uniform continuity $\int |u(x+\beta) - u(x)|\lambda(dx)$ tends to zero as β does. Thus taking a limit we obtain

$$\int H(\theta, \xi)\mu(d\theta) = S^{-\alpha} \int H(\theta, \alpha + \xi)\mu(d\theta).$$

This implies that

$$H(\theta, \xi) = S^{-\alpha} H(\theta, \alpha + \xi)$$

for all θ except those of a null set that might depend on α and ξ. However the δ_n–tail continuity implies that the $H(\theta, \xi)$ can be taken continuous in ξ. Thus approximating the set of all (α, ξ) by a dense countable set one sees that there will be one null set N such that if $\theta \notin N$ one must have $H(\theta, \xi) = S^{-\alpha} H(\theta, \alpha + \xi)$ for all (α, ξ). This completes the proof of the proposition. □

Translated back in terms of the original distributions, this means that, except for a set of Lebesgue measure zero in θ, we shall have, for weak convergences

$$\lim \mathcal{L}\{[\delta_n^{-1}(T_n - \theta), K_n] | P_{\theta + \delta_n \xi, n}\}$$

$$= \lim \mathcal{L}\{[\delta_n^{-1}(T_n - \theta - \delta_n \alpha), K_n] | P_{\theta + \delta_n(\alpha + \xi), n}\}$$

for all α and ξ. Taking $\xi = 0$ gives the following invariance property.

The limit of

$$\mathcal{L}[\delta_n^{-1}(T_n - \theta - \delta_n\alpha), K_n]|P_{\theta+\delta_n\alpha,n}\}$$

is the same as that of

$$\mathcal{L}\{[\delta_n^{-1}(T_n - \theta), K_n]|P_{\theta,n}\}.$$

In particular, the limit distribution of K_n does not depend on α. We have argued so far as if Θ was the entire space $\Re^k$. That condition can be removed, or at least weakened. It is easy to extend the result to the case where Θ is only a measurable subset of $\Re^k$. For instance, one can complete the definition of $\mathcal{E}_n = \{P_{\theta,n}; \theta \in \Re^k\}$ by letting $P_{\theta,n} = \mathcal{N}(\theta, \delta_n I)$ for $\theta \in \Re^k \setminus \Theta$.

The case where δ_n is a matrix does not differ much from the preceding. The case where δ_n is a function of θ requires a bit more care. We shall leave it alone.

Proposition 5 uses a measurability restriction. If the sequence $\mathcal{E}_n$ is δ_n–tail continuous, one can enforce such measurability without much trouble. Indeed we have the following result.

Proposition 6. *Assume that $\mathcal{E}_n = \{P_{\theta,n}; \theta \in \Re^k\}$ is δ_n–tail continuous at each θ. Then there exist other experiments $\{Q_{\theta,n}; \theta \in \Re^k\}$ such that:*

(i) the maps $\theta \rightsquigarrow Q_{\theta,n}$ are continuous for the L_1-norm, and

(ii) for each θ and each fixed $b < \infty$

$$\sup_{|t-\theta|\leq\delta_n b} \|P_{t,n} - Q_{t,n}\| \to 0.$$

Proof. Pave $\Re^k$ by cubes of diameter δ_n^2. Let $\{C_{j,n}\}$ be the jth cube of the paving. Take an open cube $C'_{j,n}$ with the same center as $C_{j,n}$ but twice the diameter. There are continuous functions $u'_{j,n}$ with $0 \leq u'_{j,n} \leq 1$, which are unity on $C_{j,n}$ and vanish outside of $C'_{j,n}$. Let $u_{j,n} = u'_{j,n}/\sum_i u'_{i,n}$. These form a continuous partition of unity. Define $Q_{\theta,n}$ by $Q_{\theta,n} = \sum_j u_{j,n}(\theta)P_{j,n}$ where

$P_{j,n} = P_{\eta,n}$ for η equal to the center of the cube $C_{j,n}$. This gives continuous maps $\theta \rightsquigarrow Q_{\theta,n}$. It is easy to check that they have the desired property. □

The limiting invariance property described in Proposition 5 also says something about the "regularity" of estimates such as T_n. This is according to the following definition.

Definition 5. *A sequence $\{T_n\}$ with T_n defined on $\mathcal{E}_n = \{P_{\theta,n}; \theta \in \Theta\}$ is called δ_n-regular at the point θ_0 if the difference*

$$\mathcal{L}\{\delta_n^{-1}(T_n - \theta_0 - \delta_n \alpha_n) | P_{\theta_0 + \delta_n \alpha_n, n}\}$$

$$-\mathcal{L}\{\delta_n^{-1}(T_n - \theta_0) | P_{\theta_0, n}\}$$

tends to zero for all bounded sequences $\{\alpha_n\}$.

According to this, we can state the following.

Proposition 7. *Let the $\mathcal{E}_n$ be δ_n–tail continuous. Then any sequence $\{T_n\}$ such that $\mathcal{L}[\delta_n^{-1}(T_n - \theta)|P_{\theta,n}]$ converges for all θ is automatically δ_n-regular at all θ except possibly those in a set of Lebesgue measure zero.*

We shall prove this at the same time as Proposition 8.

Another consequence of Proposition 5 is that the LAQ conditions often imply the stronger LAMN conditions.

Let us recall that a sequence $\mathcal{E}_n = \{P_{\theta,n}; \theta \in \Theta\}$ satisfies the LAMN condition at a point $\theta \in \Theta$ if

(i) it satisfies LAQ at θ for a sequence $\delta_n > 0$, $\delta_n \to 0$, and

(ii) if any subsequence $\{m\} \subset \{n\}$ is such that $\mathcal{L}[K_m|P_{\theta,m}]$ converges to, say, $\mathcal{L}(K)$, then $\mathcal{L}[K_m|P_{\theta+\delta_m\xi_m,m}] \to \mathcal{L}(K)$ for every bounded sequence $\{\xi_m\}$.

Proposition 8. *Assume that $\mathcal{E}_n$ satisfies LAQ at all $\theta \in \Theta$ for a given sequence $\{\delta_n\}$. Assume also that $\mathcal{L}[K_n|P_{\theta,n}]$ has a limit $\mathcal{L}(K|\theta)$ for each θ. Then $\mathcal{E}_n$ satisfies LAMN at Lebesgue almost all θ.*

Proof. Let us first note that the LAQ assumption at a point θ implies that the $\mathcal{E}_n$ are δ_n–tail continuous at that point. This follows from the fact that if $\{\xi_n\}$ and $\{\tau_n\}$ are two bounded sequences such that $\xi_n - \tau_n \to 0$, the differences

$$\Lambda_n(\theta + \delta_n \xi_n; \theta) - \Lambda_n(\theta + \delta_n \tau_n; \theta)$$

tend to zero in $P_{\theta,n}$ probability, according to part (ii) of the LAQ conditions of Section 6.2. By contiguity, the convergence also occurs in $P_{\theta+\delta_n\xi_n,n}$ probability and the same applies to τ_n. That the difference of the measures $P_{\theta+\delta_n\xi_n.n} - P_{\theta+\delta_n\tau_n,n}$ converges in norm to zero is then a consequence of elementary inequalities, recorded in Le Cam [1986, chapter 4, section 3]. Inequalities in a reverse direction can also be found there.

Now take a probability measure μ equivalent to the Lebesgue measure and a particular ξ. The condition that $\mathcal{L}[K_n|P_{\theta,n}]$ tends to a limit for each θ implies that $\mathcal{L}(K_n|P_{\theta,n})$ tends to a limit almost surely for μ, hence also for the shifted measures $S^{\delta_n\xi}\mu$. This implies that $\mathcal{L}[K_n|P_{\theta+\delta_n\xi,n}]$ converges to a limit almost surely μ for each fixed ξ. Thus the difference $\mathcal{L}[K_n|P_{\theta+\delta_n\xi,n}] -\mathcal{L}(K_n|P_{\theta,n}) = \Delta_n(\theta,\xi)$ converges to a limit almost surely μ for each ξ. This is true for a dense countable set of values of ξ. Thus, by the δ_n–tail continuity we can say that there is one set N of μ measure zero such that for $\theta \notin N$, $\Delta_n(\theta,\xi)$ tends to a limit for all ξ.

According to Proposition 6, one may as well assume that $\mathcal{L}(K_n|\theta)$ is a continuous function of θ, hence at least measurable. According to Egorov's theorem, the almost sure convergence of $\mathcal{L}(K_n|\theta)$ implies the existence of a sequence of compacts, say, C_r, $r = 1, 2 \ldots$, such that $\Re^k \setminus \cup_r C_r$ has μ measure zero and such that $\mathcal{L}(K_n|\theta)$ converges uniformly on each C_r.

Consider then a point θ that is a point of Lebesgue density unity of C_r. Let μ_n be the Lebesgue measure restricted to a ball $B_n = \{\eta : |\eta - \theta| < \delta_n b\}$ centered at θ, having radius $\delta_n b$, and renormalized to be a probability measure. The uniform convergence on C_r implies that $\mathcal{L}(K_n|\eta)$ converges in μ_n probability on B_n. However, by the assumed δ_n–tail continuity, this implies that $\mathcal{L}[K_n|\theta + \delta_n\xi]$ converges uniformly for $|\xi| \leq b$.

Since points of Lebesgue density unity are all of C_r, except a set of Lebesgue measure zero, this means that for almost all $\theta \in \Re^k$ the distributions $\mathcal{L}[K_n|\theta+\delta_n\xi]$ converge to a limit uniformly on bounded subsets of ξ. Thus the same is true of $\mathcal{L}(K_n|\theta + \delta_n\xi) - \mathcal{L}(K_n|\theta)$. However, by Proposition 5, this limit is almost everywhere zero. Hence the result.

The proof of Proposition 7 can be carried out in the same manner. The notation is somewhat more complicated, but the main argument is the same. □

According to such arguments, assumptions that sequences of experiments are LAQ will often imply that they satisfy LAMN. This does not mean that they are *always* LAMN. Jeganathan [1988] points out that, in some time series situations, the set where LAQ holds but LAMN fails is often very interesting.

There are other results that imply that some approximation conditions often have unintended consequences almost everywhere. Here is an example:

Proposition 9. *Consider a sequence $\{\mathcal{E}_n\}$, $\mathcal{E}_n = \{P_{\theta,n}; \theta \in \Re^k\}$ where at each θ the logarithm Λ_n admits an approximation such that*

$$\Lambda_n(\theta + \delta_n t_n; \theta) - [t_n' S_n - A(\theta, t_n)]$$

tends to zero for a fixed non-random *function $(\theta, t) \rightsquigarrow A(\theta, t)$ and for all bounded sequences $\{t_n\}$. Assume also that $\{P_{\theta+\delta_n t_n,n}\}$ and $\{P_{\theta,n}\}$ are contiguous for such bounded sequences $\{t_n\}$.*

Then, for almost all θ, the experiments $\mathcal{F}_{\theta,n} = \{P_{\theta+\delta_n t,n}; t \in \Re^k\}$ tend weakly to limits, say, $\mathcal{F}_\theta = \{Q_{\theta,t}; t \in \Re^k\}$, such that

$$\begin{aligned} \Lambda(t,\theta) &= \log \frac{dQ_{\theta,t}}{dQ_{\theta,0}} \\ &= t'S_\theta - \frac{1}{2} t' M_\theta t \end{aligned}$$

for some random vector S_θ and some nonrandom matrix M_θ.

Proof. By the argument of Proposition 5, such a convergence and a measurability restriction on the $P_{\theta,n}$ will imply that for almost all θ the experiment $\mathcal{F}_\theta$ is shift-invariant in the sense that for any finite set $\{t_j; j = 1, 2, \dots, m\}$ the distribution of $\{\Lambda(s + t_j; \theta); j = 1, 2, \dots, m\}$ under $Q_{\theta,s}$ is the same as that of $\{\Lambda(t_j; \theta); j = 1, 2, \dots, m\}$ under $Q_{\theta,0}$. Taking second differences

$$\Lambda(s+t,\theta) + \Lambda(s-t,\theta) - 2\Lambda(s,\theta)$$

$$= A(\theta, s+t) + A(\theta, s-t) - 2A(s,\theta),$$

one sees that

$$A(\theta, s+t) + A(\theta, s-t) - 2A(\theta, s)$$

$$= A(\theta,t) + A(\theta,-t) - 2A(\theta,0).$$

This implies that $A(\theta, t)$ is a quadratic function of t and that the $\mathcal{E}_n$ satisfy the LAN conditions, as we shall now show.

The fact that A is quadratic in t can be seen as follows. Drop the variable θ for simplicity. Then the relation becomes $A(s+t) + A(s-t) - 2A(s) = A(t) + A(-t) - 2A(0)$, with $A(0) = 0$. The relation

$$\int e^{tS - A(\theta,t)} dQ_{\theta,0} = 1$$

implies also that A is continuous in t. One can write $A(t) = V(t) + W(t)$ where $V(t) = [A(t) + A(-t)]/2$, $W(t) = [A(t) - A(-t)]/2$. It is easily checked that V has constant second differences and that W has second differences equal to zero. For V, consider the function $B(t,u) = V(t+u) - V(t) - V(u)$ for a fixed vector u. Computing second differences one sees that the function of t so obtained also has second differences equal to zero. Furthermore $W(0) = 0$ and $B(0,u) = B(t,0) = 0$. Keeping u constant we obtain

$$B(t+v,u) + B(t-v,u) - 2B(t,u) = 0$$

for every vector v. Taking $v = t$ gives the relation $B(2t, u) = 2B(t,u)$. Proceeding inductively we obtain $B(mt, u) = mB(t, u)$ for every integer m. This relation for integers implies its validity for rational numbers. Thus B must be linear in t of the form $B(t,u) = \Gamma(u)t$ where Γ is a certain matrix. However, B is symmetric: $B(t,u) = B(u,t)$. Thus $B(t,u)$ must be of the form

$u'Mt$ for some matrix M. A similar argument shows that W is linear in t. Taking $u = -t$ in $B(t,u)$ gives

$$B(t,-t) = -t'Mt = V(0) - V(t) - V(-t)$$

$$= -2V(t), \text{ because } V(t) = V(-t).$$

This gives the form of $A(t)$ as a linear term plus a quadratic term. □

Here we have used a *nonrandom* function A of θ and t. Jeganathan [1983] extended this result to the case where the local experiments $\mathcal{F}_{\theta,n} = \{P_{\theta+\delta_n t,n}; t \in \Re^k\}$ have logarithms of likelihood ratios that admit approximations of the form

$$\Lambda_n(\theta + \delta_n t; \theta) \sim t'S_{\theta,n} - A_n(\theta, t)$$

where A_n is random. He assumes, among other things, that the functions A_n are not changed much by a small change in the parameter θ.

One can then show that the A_n can be replaced by quadratic functions of t.

What is all this trying to say? It means that if one has approximations to the logarithms of likelihood ratios that start by a linear-quadratic term in t and are otherwise not too sensitive to small modifications of θ, then they will be mostly reduced to the linear-quadratic type in t. This means that the conditions LAQ will often hold and therefore that the LAMN conditions will often hold. This should be enough incentive to study the LAN and LAMN conditions more closely, which we shall now do.

6.6 The LAMN and LAN Conditions

As in previous sections of this chapter we shall consider experiments $\mathcal{E}_n = \{P_{\theta,n}\}$ and assume that Θ is the whole of $\Re^k$ or at least that θ is an interior point of $\Theta \subset \Re^k$. We shall also assume that $\{\delta_n\}$ is a given sequence of numbers, $\delta_n > 0$, $\delta_n \to 0$ as n tends to infinity.

We first restate the LAMN conditions in two equivalent forms. Then we will prove an asymptotic minimax theorem for the centerings Z_n described in Section 6.3. This is followed by a convolution theorem stated and proved only for the LAN case. The extension to a conditional convolution theorem valid for the LAMN case is not given, but it follows from the argument used in the LAN case.

Now to restate the LAMN conditions. The first condition is a contiguity condition:

(C1) The sequences $\{P_{\theta+\delta_n t_n,n}\}$ and $\{P_{\theta,n}\}$ are contiguous if t_n stays bounded.

The second condition is the same as in LAQ:

(C2) There are random vectors S_n and random symmetric matrices K_n, (usually depending on θ) such that

$$\Lambda_n(\theta+\delta_n t_n, \theta) - [t_n' S_n - \frac{1}{2} t_n' K_n t_n]$$

tends to zero in probability for every bounded sequence $\{t_n\}$.

The special condition imposed in LAMN and not in the more general LAQ is:

(C3) If a subsequence $\{m\} \subset \{n\}$ is such that $\mathcal{L}\{K_m|P_{\theta,m}\}$ converges to a limit $\mathcal{L}(K)$, then $\mathcal{L}[K_m|P_{\theta+\delta_m t_m,m}]$ converges to that same limit $\mathcal{L}(K)$ for any bounded sequence $\{t_m\}$.

Condition (C3) can be replaced by another one that looks somewhat different.

(C4) If a subsequence $\{m\} \subset \{n\}$ is such that $\mathcal{L}[K_m|P_{\theta,m}]$ converges to a limit $\mathcal{L}(K)$, then $\mathcal{L}\{S_m, K_m|P_{\theta,m}\}$ converges to a limit $\mathcal{L}(S,K)$, where conditionally given K, the vector S is gaussian with $E(S|K)=0$ and $E[(t'S)^2|K]=t'Kt$.

Lemma 3. *Under the other conditions imposed here, conditions (C3) and (C4) are equivalent.*

Proof. We already know, by Section 6.2, that K must be almost surely positive semidefinite. Taking a further subsequence if necessary, we can assume that $\mathcal{L}[S_m, K_m|P_{\theta,m}]$ tends to a limit $\mathcal{L}(S,K)$, as discussed in Section 6.2. We first show that (C3) implies (C4). Let γ be any bounded continuous function on the space of symmetric matrices. To pass to the limit under $P_{\theta+\delta_m t,m}$, use the contiguity condition and Section 3.1, Proposition 1. The fact that $\mathcal{L}(K)$ does not depend on the value of t for $P_{\theta+\delta_m t,m}$ is seen to mean that, taking expectations under the limit for $P_{\theta,m}$,

$$E\gamma(K)e^{t'S-\frac{1}{2}t'Kt} \equiv E\gamma(K).$$

Letting

$$\psi(K,t) = E[e^{t'S-\frac{1}{2}t'Kt}|K],$$

this can be written

$$E\gamma(K)\psi(K,t) = E\gamma(K)$$

for all bounded continuous functions γ. It follows that $\psi(K,t) = 1$ almost surely for each t. The relation $E[e^{t'S-\frac{1}{2}t'Kt}|K] = 1$ almost surely implies that, given K, the vector S is gaussian with mean zero and covariance given by $E[(t'S)^2|K] = t'Kt$.

Conversely, to show that (C4) implies (C3), note that if $E[e^{t'S-\frac{1}{2}t'Kt}|K] = 1$ almost surely, the expectation of $\gamma(K)$ for the limit measure at t is independent of t. □

The result of Lemma 3 is well-known. It was observed independently and almost simultaneously by Davies, Jeganathan and Swensen in 1980.

Lemma 3 refers to the LAMN conditions. One of its important implications is that a stability condition on the limit law of K_n implies *conditional* asymptotic normality of S_n. This applies, of course, also to the LAN conditions, but there the limits of the K_n are not random. So the conditioning is unnecessary. Although this is a consequence of Lemma 3, we shall give a separate proof because it is much simpler.

Lemma 4. *Let the LAN conditions be satisfied at θ. That is, the conditions (O) and (C) of Section 6.2 are satisfied. For bounded sequences t_n, the difference*

$$\Lambda_n(\theta+\delta_n t_n,\theta) - [t_n' S_n - \frac{1}{2}t_n' K_n t_n]$$

tends to zero in $P_{\theta,n}$ probability, and the cluster points of K_n are not random. Then S_n is asymptotically normally distributed with covariance matrix K_n.

Proof. Take a subsequence along which $\mathcal{L}(S_n, K_n)$ has a limit, say, $\mathcal{L}(S, K)$. The contiguity condition implies (see Chapter 3) that $E \exp\{tS - (1/2)t'Kt\} = 1$. Since K is not random, this reads $E \exp(tS) = \exp(t'Kt/2)$. One recognizes the formula for the Laplace transform of a gaussian random vector. The result follows since condition (O) implies that the formula holds for all t.□

Although it is possible to proceed under conditions (C1), (C2), and (C3) (or (C4)) we shall add, as in LAQ, a non-degeneracy clause:

(C5). If $\mathcal{L}[K_m | P_{\theta,m}]$ is a subsequence converging to a limit $\mathcal{L}(K)$, then K is almost surely invertible.

Lemma 3 can then be restated by saying that under (C1) to (C5), if $\mathcal{L}[K_m | P_{\theta,m}]$ converges to $\mathcal{L}(K)$, then $\mathcal{L}[S_m, K_m | P_{\theta,m}]$ converges to

$$\mathcal{L}(K^{1/2} Z, K)$$

where Z is $\mathcal{N}(0, I)$ independent of K.

We are now in a position to state a local asymptotic minimax theorem as Theorem 1, but already essentially part of Lemma 5.

Consider the local experiments $\mathcal{F}_{\theta,n} = \{P_{\theta+\delta_n \tau, n}; \tau \in \Re^k\}$. Give to τ a prior distribution that is gaussian centered at zero with inverse covariance matrix Γ. Consider the problem of estimating τ by $\hat{\tau}$ with a loss function $W(\hat{\tau}, \tau) = W(\hat{\tau} - \tau)$ with W "bowl shape" that is the sets $\{u : W(u) \leq \alpha\}$ are closed convex symmetric sets.

Lemma 5. *Assume that (C1) to (C5) are satisfied and that for a sequence $\{m\} \subset \{n\}$ the distributions $\mathcal{L}[K_m | P_{\theta,m}]$ have a limit $\mathcal{L}(K)$. Then, for any bounded bowl shape loss function, W*

$$\lim_{\Gamma \to 0} \lim_{m} \inf_{\hat{\tau}_m} E_m W(\hat{\tau}_m - \tau) = E W(K^{-1/2} Z)$$

where the first expectation E_m is taken for the joint distribution induced by the $P_{\theta+\delta_m\tau,m}$ and the gaussian distribution of τ and where, in the second expectation, Z is $\mathcal{N}(0,I)$ independent of K.

Proof. We already saw in Section 6.4, Proposition 4, that, for bounded W, the posterior distribution of τ can be replaced by a normal one centered at a point $T_n = \theta + (K_n + \Gamma)^{-1}K_n(Z_n - \theta)$ and having $(K_n + \Gamma)$ as inverse covariance matrix. For all bowl shape loss functions, Anderson's lemma (see Section 6.4) says that the minimum of the posterior risk for the normal approximation is achieved at T_n. It is then equal to the conditional expectation $E\{W[(K_n + \Gamma)^{-1/2}Z]|\mathcal{B}_n\}$ where Z is $\mathcal{N}(0,I)$ and $\mathcal{B}_n$ is the σ-field of the observations. This can be replaced by $E\{W[(K_n + \Gamma)^{-1/2}Z]|K_n\}$.

The actual Bayes risk is obtained by integrating this with respect to the marginal distribution of K_n.

Now take a subsequence $\{m\} \subset \{n\}$ such that $\mathcal{L}[K_m|P_{\theta,m}] \to \mathcal{L}(K)$. Then, by Lemma 3, $\mathcal{L}[K_m|P_{\theta+\delta_m\tau,m}]$ also tends to $\mathcal{L}(K)$. Thus in our bayesian system, the marginal distribution of K_m also tends to $\mathcal{L}(K)$. Thus the joint distribution $\mathcal{L}(Z_m, K_m)$ tends to that of (Z, K) with Z independent of K and $\mathcal{N}(0,I)$. This gives the limit $EW[(K+\Gamma)^{-1/2}Z]$ for the risk. The stated result follows if one lets Γ tend to zero since $W(K^{-1/2}Z)$ has a set of discontinuities of measure zero. □

As a corollary one can state the following local asymptotic minimax theorem.

Theorem 1. *Let (C1) to (C5) be satisfied and let W be a non-negative bowl shape loss function. Assume $\mathcal{L}[K_n|P_{\theta,n}] \to \mathcal{L}(K)$. Then for any sequence of estimates T_n of τ, one has*

$$\lim_{b\to\infty} \lim_{c\to\infty} \liminf_n \sup_{|\tau|\leq c} E\{b \wedge W(T_n - \tau)|\theta + \delta_n\tau\}$$

$$\geq EW(K^{-1/2}Z)$$

where Z is $\mathcal{N}(0,I)$ independent of K. Under the conditions on auxiliary estimates of Section 6.3, the lower bound is achieved *by* $T_n = \delta_n^{-1}(Z_n - \theta)$ *where* Z_n *is the centering* Z_n *of that section.*

Proof. Lemma 5 applies to the bounded function $b \wedge W$. Take a given Γ positive definite and a c so large that the normal prior probability of $|\tau| > c$ is at most ϵ/b. Then

$$\sup_{|\tau| \le c} E\{b \wedge W(T_n - \tau)|\theta + \delta_n \tau\}$$

is at least $E[b \wedge W(T_n - \tau)] - \epsilon$ where the expectation is taken over T_n and τ. Thus, for such a c, we have

$$\liminf_n \sup_{|\tau| \le c} E[b \wedge W(T_n - \tau)|\theta + \delta_n \tau]$$

$$\ge E\{b \wedge W[(K + \Gamma)^{-1/2} Z]\} - \epsilon.$$

Letting c tend to infinity gives a limit at least equal to $Eb \wedge W[(K + \Gamma)^{-1/2} Z]$ for all Γ. Then the result follows by letting Γ tends to zero and b tend to infinity. □

We have stated this local asymptotic minimax theorem in the usual form, imitated from Hájek [1972]. However, it is not the best possible form. There is a theorem of Le Cam [1979], (see also Le Cam [1986, p. 109]) that says the following.

Proposition 10. *Let* Ξ *be a fixed set and let* W *be a fixed loss function. Let* $\mathcal{F}_n$ *be experiments indexed by* Ξ *and converging weakly to a limit* $\mathcal{F}$. *Suppose that* W *is bounded from below for each* $\xi \in \Xi$. *Let* $\bar{\mathcal{R}}(\mathcal{F})$ *be the augmented closed set of risk functions defined in Chapter 2. Define* $\bar{\mathcal{R}}(\mathcal{F}_n|S)$ *similarly, for a subset* S *of* Ξ *and for experiments restricted to* S *and risk functions obtainable on* S.

Suppose that r *is a function that does not belong to* $\bar{\mathcal{R}}(\mathcal{F})$. *Then there are numbers* $a > 0, b > 0$, *a finite set* $S \subset \Xi$ *and an integer* N *such that if* $n \ge N$, *then* $(r \wedge b) + a$ *restricted to* S *does not belong to* $\bar{\mathcal{R}}(\mathcal{F}_n|S)$.

The proof is not difficult, but it requires some preparation. If applied here it can give the following.

Theorem 2. *Let (C1) to (C5) be satisfied and let W be a non-negative bowl shaped function. Assume $\mathcal{L}[K_n|P_{\theta,n}] \to \mathcal{L}(K)$. Let a be any number $a < EW(K^{-1/2}Z)$. Then there is a finite set $F = \{\tau_1, \ldots, \tau_m\}$, a number $b < \infty$, and an integer N such that $n \geq N$ implies*

$$\inf_{T_n} \sup_{\tau \in F} E\{b \wedge W(T_n - \tau)|\theta + \delta_n \tau\} > a.$$

Proof. The experiments $\mathcal{F}_{\theta,n} = \{P_{\theta+\delta_n\tau,n}; \tau \in \Re^k\}$ converge weakly (see Chapter 2) to an experiment $\mathcal{F} = \{Q_\tau; \tau \in \Re^k\}$, where

$$\log \frac{dQ_\tau}{dQ_0} = \tau' S - \frac{1}{2}\tau' K \tau$$

with a random vector S and a random positive-definite matrix K such that, given K, the conditional distribution of S is $\mathcal{N}(0, K)$.

Let $\mathcal{R}(\mathcal{F}, W)$ be the set of functions that are risk functions or are larger than risk functions for this experiment $\mathcal{F}$ and the loss function W. The system used in Le Cam [1979, 1986] involves "transitions" that are not of Markov kernels but limits of them and a definition of risk such that $\mathcal{R}(\mathcal{F}, W)$ is always equal to its pointwise closure $\bar{\mathcal{R}}(\mathcal{F}, W)$. This is needed for the proof, but only for the limit experiment $\mathcal{F}$. In fact, to apply the argument here all one needs to show is that the function identically equal to $a < EW(K^{-1/2}Z)$ is not in the closed set $\bar{\mathcal{R}}(\mathcal{F}, W)$.

To show this, put on $\Re^k$ a gaussian $\mathcal{N}(0, \Gamma^{-1})$ prior. For the limit experiment $\mathcal{F}$ one obtains a gaussian posterior distribution with covariance matrix $(K + \Gamma)^{-1}$. The argument, using Anderson's lemma, that gives the posterior risk does not depend on whether one uses genuine Markov kernels or limits of them. Thus, calling μ the $\mathcal{N}(0, \Gamma^{-1})$ measure, one will have

$$\inf_f \{\int f d\mu; f \in \bar{\mathcal{R}}(\mathcal{F}, W)\} = EW[(K + \Gamma)^{-1/2} Z].$$

However, if $a < EW(K^{-1/2}Z)$, taking Γ sufficiently small will yield $a < EW[(K + \Gamma)^{-1/2}Z]$. This is incompatible with $a \in \bar{\mathcal{R}}(\mathcal{F}, W)$. Hence the result. □

Note that the method of proof of Theorem 2 involves a Bayes argument but only on the limit experiment $\mathcal{F}$. The approximation of posterior distributions by gaussian ones described in Section 6.4 and used in the proof of Theorem 1 is not involved here.

Note also that the truncation level b and the finite set F of Theorem 2 depend only on the loss function W, the choice of the number a, and the limiting distribution $\mathcal{L}(K)$ which characterizes the limit experiment $\mathcal{F}$. They do not depend on other features of the sequence $\{\mathcal{F}_{\theta,n}\}$. The latter is involved only in that its convergence to $\mathcal{F}$ when restricted to F will yield the number N.

Remark 11. If the sets $\{u : W(u) \leq \alpha\}$ for $\alpha < \sup_u W(u)$ are compact, it is easy to show that $\mathcal{R}(\mathcal{F}, W)$ is already closed; see Le Cam [1955]. It should be pointed out that, under (C1) to (C5) and the assumptions of Proposition 1, Section 6.3, *the centerings* Z_n *constructed in Section 6.3 yield estimates* $T_n = \delta_n^{-1}(Z_n - \theta)$ *of* τ *that will achieve the minimax risk* $EW(K^{-1/2}Z)$ *in the limit.*

Remark 12. Theorem 2 is stronger than Theorem 1 in the sense that the finite set F, the number b, and the integer N are independent of the sequence $\{T_n\}$.

Another theorem that can be used to show that these centerings have good asymptotic properties is Hájek's convolution theorem. To simplify, we shall state it only for the LAN case. The statement and proofs for the LAMN case are similar. As Jeganathan [1983] showed, the convolution theorem is valid conditionally, given the limit matrix K.

Theorem 3. (Hájek) *Let conditions (C1) and (C2) be satisfied for matrices* K_n *such that* $\mathcal{L}[K_n|P_{\theta,n}]$ *tends to* $\mathcal{L}(K)$ *for a nonrandom positive-definite matrix* K. *For the experiments* $\mathcal{F}_{\theta,n} = \{P_{\theta+\delta_n\tau,n}; \tau \in \Re^k\}$, *let* T_n *be an estimate of* $A\tau$ *where* A *is a given (nonrandom) matrix. Assume that* $\mathcal{L}[T_n - A\tau|\theta + \delta_n\tau]$ *tends to a limit* H *independent of* τ. *Then* H *is the distribution of* $AK^{-1/2}Z + U$ *where* Z *is* $\mathcal{N}(0, I)$ *independent of the random vector* U.

Proof. (van der Vaart). Consider the approximations $\tau' S_n - (1/2)\tau' K_n \tau$ to the logarithms $\Lambda_n(\theta + \delta_n \tau; \theta)$. Since $K_n \to K$, one can substitute K for K_n in that formula. It will be more convenient to write it in the form $\tau' S_n - (1/2)\tau' K \tau = \tau' K X_n - (1/2)\tau' K \tau$ where $X_n = K^{-1} S_n$.

Taking a subsequence if necessary, one can assume that under $P_{\theta,n}$, the joint distribution of (X_n, T_n) tends to a limit L, distribution of a pair (X, T). Then by contiguity (see Chapter 3, Proposition 1) under $\theta + \delta_n \tau$, the joint distribution of (X_n, T_n) tends to a limit that can be written in the form

$$L(dx, dt|\tau) = \exp\{\tau' K x - \frac{1}{2}\tau' K \tau\} L(dx, dt).$$

Now give τ a normal $\mathcal{N}(0, \Gamma^{-1})$ distribution. The joint distribution becomes

$$L(dx, dt, d\tau) = C_1 \exp\{-\frac{1}{2}\tau'(K + \Gamma)\tau + \tau' K x\} L(dx, dt) d\tau$$

where C_1 is the usual constant $|\det \Gamma/(2\pi)^k|^{1/2}$.

Introduce new variables $\xi = \tau - (K + \Gamma)^{-1} K x$ and $y = t - A(K + \Gamma)^{-1} K x$. To simplify notation, let $B = (K + \Gamma)^{-1} K$ and $V = BX$. Rewriting $(1/2)\tau'(K + \Gamma)\tau - \tau' K x$ in the form

$$\frac{1}{2}(\tau - v)'(K + \Gamma)(\tau - v) + \frac{1}{2} v'(K + \Gamma) v,$$

one sees that conditionally given x and t, or equivalently given v and y, the distribution of $\xi = \tau - v$ is a centered gaussian distribution G_Γ with inverse covariance matrix $(K + \Gamma)$. Thus the joint distribution of the variables ξ, y, and v could be written $G_\Gamma(d\xi) M(dy, dv)$ where M is the marginal distribution of $Y = T - ABX$ and $V = BX$. That is, ξ and (Y, V) are independent. Now note that

$$(T - A\tau) = (T - ABX) - A(\tau - BX) = Y - A\xi.$$

Thus the distribution of $T - A\tau$, in the joint distribution of X, T, and τ, is the same as that of $Y - A\xi$ where ξ and Y are independent and ξ has distribution G_Γ. Note that we don't have to bother about X itself but only $T - A\tau$ on one side and Y and ξ on the other.

So far we have not used any assumption about the limiting distribution $\mathcal{L}[T - A\tau|\tau]$. The theorem assumes that it is independent of τ and equal to some measure H. The marginal of $T - A\tau$ in the joint distribution $\mathcal{L}(\tau, T - A\tau)$ is then H itself. Thus, under the assumption, we see that H is the convolution of the gaussian distribution of $A\xi$ with some other distribution, namely, that of Y.

Now let Γ tend to zero. This does not affect $H = \mathcal{L}(T - A\tau|\tau)$ independent of τ. The gaussian distribution of ξ that was $\mathcal{N}[0, (K + \Gamma)^{-1}]$ tends to $G = \mathcal{N}(0, K^{-1})$. The distribution of Y tends to that of $T - AX$. Thus if that distribution is μ we have $H = G * \mu$. That would prove the theorem except for the circumstance that we have argued on a subsequence where the joint distribution of (X_n, T_n) tends to a limit. However, G is well determined and the limit H is assumed to exist. Thus μ is also well determined. Hence the result. □

This theorem can be applied in particular when A is the identity map of $\Re^k$ into itself. It then says that the limiting distribution of any "regular" sequence of estimates of τ must be $\mathcal{N}(0, K^{-1})$ convoluted with some other probability measure. Therefore it is less concentrated than $\mathcal{N}(0, K^{-1})$.

Under the circumstances of Proposition 1, Section 6.3, we obtained centering sequences $\delta_n^{-1}(Z_n - \theta)$ that have the $\mathcal{N}(0, K^{-1})$ limiting distribution. Thus, these centerings could be called "asymptotically most concentrated."

Remark 13. We have given the statement of Theorem 3 only in the LAN case where the limit K is nonrandom. The LAMN case can be treated similarly, conditioning all the distributions in sight on a given value for K.

Remark 14. The argument of Theorem 3 can be carried out with inessential modifications for the case where the parameter τ takes its value in an infinite-dimensional space. The modifications involve notational conventions and not much else, especially if one assumes that A is a map into a finite-dimensional space or that the limiting distribution of $T_n - A\tau$ is tight. To indicate the possibilities, consider the case where τ takes its values in some arbitrary linear space, say, F. A gaussian shift experiment indexed linearly by F is then given by mutually absolutely continuous measures Q_τ such that $\log(dQ_\tau/dQ_0) = \langle \tau, X\rangle - (1/2)\|\tau\|^2$ where, under Q_0, the symbol $\langle \tau, X\rangle$ means evaluation at τ of a certain linear gaussian process with $E_0\langle \tau, X\rangle = 0$ and $E_0|\langle \tau, X\rangle|^2 = \|\tau\|^2$. Here $\|\bullet\|$ is a certain pre-hilbertian pseudonorm. Now suppose that our local experiments $\mathcal{F}_{\theta,n} = \{P_{\theta+\delta_n\tau,n}; \tau \in F\}$ tend weakly to the gaussian limit $\mathcal{G} = \{Q_\tau; \tau \in F\}$.

Officially $\|\bullet\|$ is only a pseudonorm. However, if $\mathcal{L}\{T_n - A\tau|\theta + \delta_n\tau\} \to H$ in any reasonable sense, then $\mathcal{L}(T_n|\theta + \delta_n\tau)$ tends to H shifted by $A\tau$. If $\|\tau\| = 0$, then the gaussian limits Q_τ and Q_0 are the same. This implies that the limits of $\{\mathcal{L}[T_n|\theta]\}$, $\{\mathcal{L}[T_n|\theta + \delta_n\tau]\}$ must be the same and therefore that H shifted by $A\tau$ is the same as H. This means that $A\tau$ must also be zero. Thus one can take a quotient space, putting equal to zero all τ's such that $\|\tau\|^2 = 0$. Then our pseudonorm becomes a norm.

The map A sends F into some other linear space, say, F_1. We shall assume that F_1 is given a Hausdorff locally convex topology and that H is a Radon measure for it. The foregoing argument then shows that A must be continuous from $\{F, \|\bullet\|\}$ to F_1. This needs proof but is easy. (Pretend that F_1 is a normed space. Find a compact C such that $H(C) \geq 3/4$. Then there is some b such that if $|A\tau_1 - A\tau_2| \geq b$ then $C + A\tau_1$ and $C + A\tau_2$ have no points in common. If A is not continuous there are $\{\tau_\nu\}$, $\|\tau_\nu\| \to 0$ such that $\|A\tau_\nu\| > b$. This implies that H shifted by $A\tau_\nu$ and H are at L_1-distance at least $1/4$. However, they are at an L_1-distance that tends to zero since their Hellinger distance is at most $1 - \exp[-\|\tau_1 - \tau_2\|^2/8]$.)

It follows that one can extend all maps to the case where $\{F, \|\bullet\|\}$ is already complete. So we can assume that this is the case. Thus we might as well assume that $\{F, \|\bullet\|\}$ is a Hilbert space.

Now take on F a gaussian measure $\mathcal{N}(0, \Gamma^{-1})$ that is a Radon measure on $\{F, \|\bullet\|\}$.

The computations carried out in the proof of Theorem 3 work without any major difficulty, and the distribution H is seen to be that of $Y - A\xi$ where Y and ξ are independent with $\mathcal{L}(Y) = \mathcal{L}(T - ABX)$ and $A\xi$ being gaussian with the distribution image of G_Γ by A. Those are Radon measures since with the hypothesized distribution of τ the element ξ has for distribution the Radon measure G_Γ. Here we have $H = \bar{G}_\Gamma * D_\Gamma$ where both $\bar{G}_\Gamma = \mathcal{L}(A\xi; \Gamma)$ and D_Γ are Radon measures on F_1. This implies that, as Γ tends to zero, their images in any of the Banach spaces associated to F_1 must form a shift-compact "net." However, since the $\bar{G}_\Gamma$ are centered at zero, their family must be relatively compact, and so will the D_Γ. Passing to the limit as $\Gamma \to 0$ we get that $H = G * \mu$ where G is the normal distribution of AX and μ is some fixed Radon measure. In other words, the result is the same as that of Theorem 3, but with a possibly infinite-dimensional vector space for the range of the parameter τ. $\square$

Results of this type were obtained, after the basic result of Hájek, by Moussatat [1976] and, later, Millar [1985]. They are applicable to the so-called "semi-parametric" situations where τ would be a parameter in some nondescript function space but $A\tau$ would take values in some nice space, such as the line, a euclidean space, or other spaces of interest. As shown by Moussatat [1976], the convolution theorem can even be extended further, using cylindrical measures. Another extension recently given by A. van der Vaart and J. Wellner concerns the situation where the "distributions" $\mathcal{L}(T_n - A\tau | \theta + \delta_n \tau)$ are not well defined because of lack of measurability properties, but where the Prohorov type of convergence is replaced by the convergence attributed by Dudley [1985] to Hoffman-Jorgensen [1985].

We have given here Aad van der Vaart's proof of the convolution theorem because it is relatively direct and copes automatically with linear functions of the parameter (the $A\tau$). There are other possible proofs; one was pointed out by Le Cam [1972].

That particular author likes it because: "It can be conveyed to a knowledgeable listener without carrying out any computations." It relies on the Markov-Kakutani fixed point theorem in a form imitated from Eberlein [1949] and on a theorem of J. G. Wendel [1952]. The application of the Markov-Kakutani theorem poses no problem in the case considered by van der Vaart since vector spaces are abelian groups for addition. Wendel's theorem is about convolution algebras of finite measures dominated by the Haar measure on locally compact groups. It would cover only the case of finite-dimensional vector spaces. Extensions that would cover Moussatat [1976] or the results of van der Vaart [1991] require further approximation arguments. In the gaussian case, those are easy.

6.7 Additional Remarks on the LAN Conditions

For simplicity let us assume that we have experiments $\mathcal{E}_n = \{P_{\theta,n}; \theta \in \Theta\}$ where Θ is the whole of $\Re^k$. Fix a θ and consider the experiments $\mathcal{F}_{\theta,n} = \{P_{\theta+\delta_n\tau,n}; \tau \in \Re^k\}$ where $\delta_n > 0$, $\delta_n \to 0$. The LAN conditions, meaning "locally asymptotically normal." seem to say that something about the $\mathcal{F}_{\theta,n}$ is close to the gaussian situation. Unfortunately, LAN is not a very precise appellation. Here it means that at θ the LAQ assumptions are satisfied with approximations to $\Lambda_n(\theta + \delta_n\tau; \theta)$ of the form $\tau' S_n - (1/2)\tau' K_n \tau$ where the K_n tend in probability to a nonrandom limit K.

One can consider other interpretations of the words "locally asymptotically normal." One possible one is:

(N1) As n tends to infinity the experiments $\mathcal{F}_{\theta,n}$ tend to a gaussian shift experiment $\mathcal{G}_\theta = \{G_{\theta,\tau}; \tau \in \Re^k\}$.

Here a gaussian shift experiment means a family $\{Q_\tau\}$ of probability measures such that:

(i) the Q_τ are mutually absolutely continuous, and

(ii) for Q_0 the process $\tau \rightsquigarrow \log(dQ_\tau/dQ_0)$ is a gaussian process.

This is as in Chapter 4, which contains more details.

In (N1) one can specify the mode of convergence in many different ways; see Chapter 2 for strong orweak convergence. The one that fits best with our previous definitions is as follows.

Take finite sets F_n that are bounded and of bounded cardinality independent of n. Then the distance between the experiments $\{P_{\theta+\delta_n\tau,n}; \tau \in F_n\}$ and a gaussian shift experiment tends to zero. (See Chapter 2 for the definition of distances between experiments.) Note that this is not weak convergence of experiments since the F_n are not fixed.

Now there are many circumstances where such convergence properties hold but the LAN conditions are not satisfied. A standard example is given by the case of $P_{\theta,n}$, joint distribution of n i.i.d. observations $X_1, \ldots, X_n$ with common density $C\exp\{-|x-\theta|^\alpha\}$, where α is fixed, $0 < \alpha < 1/2$. Here one can take $\delta_n = n^{-\beta}$, $\beta = (1+2\alpha)^{-1}$. The experiments $\{P_{\theta+\delta_n\tau,n}; \tau \in \Re\}$ tend to a gaussian shift $\mathcal{G} = \{Q_\tau; \tau \in \Re\}$. However, if $\Lambda(\tau) = \log(dQ_\tau/dQ_0)$, the covariance under Q_0 of $\Lambda(\tau_1)$ and $\Lambda(\tau_2)$ is not bilinear in (τ_1, τ_2), as required by the LAN conditions. In fact, it is the covariance of a centered gaussian process $\{X(t); t \in \Re^k\}$ where $E|X(s) - X(t)|^2 = |s-t|^\beta$; see Section 7.3.

The LAN conditions do not allow convergence to such a limit experiment. In fact they are equivalent to the following.

Condition (N1) holds with the interpretation described earlier for convergence to a gaussian limit experiment $\mathcal{F}_\theta = \{Q_{\theta,\tau}; \tau \in \Re^k\}$. In addition, the gaussian process $X(\tau) = \log(dQ_{\theta,\tau}/dQ_{\theta,0})$ has a covariance kernel given by $C(s,t) = s'Kt$, where K is a certain positive-definite matrix.

In other words, beside the fact that the process $\log(dQ_{\theta,t}/dQ_{\theta,0})$ is gaussian, LAN implies that it is of the form $t'S - (1/2)t'Kt$ for a certain matrix K and a certain normal $\mathcal{N}(0,K)$ vector S. The linear structure of $\Re^k$ is then reflected in the form of the logarithms of likelihood ratios and in the form of the covariance kernels.

This special role for the vector space structure of $\Re^k$ is also reflected in the formulation of the LAMN and even the LAQ conditions. In a more reasonable system of assumptions, there would be no reason to expect that the linear structure of $\Re^k$ would play such an important role. For instance, one could conceive of approximations of the type

$$\Lambda(\theta + \delta_n t; \theta) \sim \xi'(t) S_n - \frac{1}{2}\xi'(t) K_n \xi(t).$$

See Section 6.9 for more elaboration.

A similar characterization could be carried out for the LAMN assumptions. The more general LAQ families present problems because properties of pairs (Z, K) for which

$$E[\exp\{t'Z - \frac{1}{2}t'Kt\}] = 1$$

are not well known.

6.8 Wald's Tests and Confidence Ellipsoids

Consider a sequence of experiments $\{\mathcal{E}_n\}$ with $\mathcal{E}_n = \{P_{\theta,n}; \theta \in \Theta\}$ as usual and a sequence $\{\delta_n\}$ of numbers $\delta_n > 0$, $\delta_n \to 0$ as n tends to infinity. For simplicity we shall assume that $\Theta = \Re^k$. We shall also assume throughout that the LAN conditions are satisfied at each θ (except for some remarks concerning the LAMN conditions).

When the LAN conditions are satisfied and when one has auxiliary estimates θ_n^* as in Section 6.3, one can form pairs (Z_n, K_n) where the Z_n's are appropriate center vectors and the matrices K_n are estimates of the inverses of the covariance matrices of the normal approximations to the distributions of Z_n.

It is then tempting to use the Z_n's as if they were actually normally distributed. For instance, one is tempted to use the quadratic $\delta_n^{-2}(Z_n - \theta)'K_n(Z_n - \theta)$ as if it had a chi-square distribution to construct tests or confidence regions for θ.

The use of such methods for one- or two-dimensional θ goes back at least to Laplace [1810a, b]. However, in 1943 Wald published a major paper in which he proved various asymptotic optimality properties for such methods. As a result, the tests so obtained are often called Wald's tests. We shall not give here a complete account of the method, but we shall indicate why and how it works.

Let us look at the local experiments $\mathcal{F}_{\theta,n} = \{P_{\theta+\delta_n\xi}; \xi \in \Re^k\}$. According to the LAN conditions these experiments tend weakly to a gaussian shift experiment $\mathcal{G}_\theta = \{Q_\xi; \xi \in \Re^k\}$ where the Q_ξ are mutually absolutely continuous such that

$$\Lambda(\xi) = \log \frac{dQ_\xi}{dQ_0} = \xi' S - \frac{1}{2}\xi' K \xi$$

where S is gaussian with expectation zero and covariance matrix K given by $E(\xi' S)^2 = \xi' K \xi$. In fact, the convergence to this gaussian shift experiment occurs uniformly on the compacts of $\xi \in \Re^k$ (for each θ).

It is a theorem (see Le Cam [1986]) that, along the sequence $\{\mathcal{F}_{\theta,n}\}$ one cannot achieve asymptotically risk functions that are not achievable on the gaussian limit.

We already referred to this result in Section 6.6, Proposition 10. Thus if one has tests, estimates, or confidence sets that are optimal for the gaussian shift case and if one can mimic them acceptably on the sequences $\{\mathcal{F}_{\theta,n}\}$, one will obtain asymptotically optimal procedures.

Rewrite the formula for $\Lambda(\xi)$ above in terms of $Z = K^{-1}S$, in the form

$$\Lambda(\xi) = -\frac{1}{2}\{(Z-\xi)' K (Z-\xi) - Z' K Z\}.$$

Wald [1943] showed that confidence ellipsoids for ξ based on the inequality $(Z-\xi)'K(Z-\xi) \le c$ where c is the appropriate α–cutoff point in a χ^2_k distribution have a certain number of optimality properties.

Under the LAN conditions these properties will be inherited, as $n \to \infty$, by the confidence sets

$$\{\theta : \delta_n^{-2}(Z_n - \theta)' K_n (Z_n - \theta) \le c\}.$$

In other words, most tests or confidence sets used for the gaussian shift case are given by sets or functions whose discontinuities have measure zero for the gaussian limits. Thus it is permissible to use the same recipes in terms of our center variables Z_n with matrices K_n. One can pass to the limit for the relevant distributions or probabilities. Furthermore, according to the principle recalled above, if one has optimality properties for risks in the gaussian shift case, they will translate automatically into asymptotic optimality properties along the sequence $\{\mathcal{E}_n\}$.

Note, However, that the LAN conditions are usually stated as conditions that are valid *pointwise* in θ. Thus the convergence and optimality properties are also valid only pointwise. If the LAN conditions hold uniformly on certain sets and the auxiliary estimates used in Section 6.3 satisfy conditions that are also appropriately uniform, then one can obtain the corresponding uniformity properties for tests on confidence sets.

The matter of practical use of the tests and confidence sets suggested by Wald is not always as simple as could be judged from the preceding. It is unfortunate that difficulties can occur as shown, for instance, by Vaeth [1985]. One type of difficulty can already be seen on the common binomial distribution $\binom{n}{k} p^k(1-p)^{n-k}$ for $p \in (0,1)$ if it is rewritten in its logistic form $\binom{n}{k} e^{\theta k}(1+e^{\theta})^{-n}$. If, for example, $n = 200$ and k happens to take small values, the Wald recipe intervals for θ become very large and their boundaries are not monotone in k. This phenomenon does not occur for the parametrization by the probability p itself, nor does it occur for the variance-stabilized $\arcsin \sqrt{p}$.

Of course, for $n = 200$, a very small k indicates that the normal theory is not applicable.

Wald's [1943] conditions were meant to imply that the entire experiments $\{P_{\theta,n}; \theta \in \Theta\}$ were approximable by heteroschedastic gaussian experiments where the log likelihood would contain terms of the type $(Z_n - \theta)'\Gamma_n(\theta)(Z_n - \theta)$ with Γ_n varying slowly enough as θ varies so that, locally, one could also approximate by gaussian shift experiments. The heteroschedastic approximation was supposed to be uniform in θ so that it would have been an approximation in the sense of the distance introduced in Chapter 2. It is clear that the binomial distribution with $p \in (0,1)$ cannot satisfy such conditions.

The matter of lack of invariance under smooth transformations of the parameters is annoying, but it is essential to the system since one deals with linear-quadratic local approximations of the type $\xi' S_n - (1/2)\xi' K_n \xi$. The choice of the parametrization will influence what is called "linear."

Attempts have been made to select "good" parametrizations that would not lead to troubles. Vaeth [1985] studied the problem of avoiding gross misbehavior for exponential families. Mammen [1987] obtained results that suggests that, in the LAN case, the local parametrization should be such that $\xi' K \xi$ is close to the quantity $-8 \log \int \sqrt{dP_{\theta,n} dP_{\theta+\delta_n \xi, n}}$.

So far, we have considered only the case covered by the LAN conditions. The LAMN case is similar but more complicated. There, in the limiting distributions $\mathcal{L}(Z, K|\xi)$ for (Z_n, K_n) under $P_{\theta+\delta_n \xi, n}$, the matrices K stay random. However, they play the role of ancillary statistics since their marginal distribution does not depend on ξ. Thus, one is tempted to condition on K and use as before tests and confidence sets based on the gaussian shift case. This is all right, but there are two additional problems. One of them can be seen most easily if ξ is taken to be one-dimensional. The expansion then looks like $\xi S - (1/2)\sigma^2 \xi^2$. Now consider the problem of testing $\xi = 0$ against $\xi > 0$. If σ^2 stays random there is no uniformly most powerful test. The test based on the normal theory given σ^2 at a fixed conditional level of significance is not the best Neyman-Pearson test of, say, $\xi = 0$ against $\xi = 1$. Swensen [1980] has shown that for best tests of $\xi = 0$ against $\xi_1 > 0$ and $\xi_2 > 0$, $\xi_1 \neq \xi_2$ to coincide, σ^2 must be nonrandom.

Another problem is that convergence of joint distributions for (Z_n, K_n) does not usually entail convergence of the conditional distributions $\mathcal{L}(Z_n|K_n)$. A discussion of the problem was given by Feigin [1986].

It can be shown that, under the LAMN assumptions, convergence of the conditional distributions $\mathcal{L}(Z_n|K_n)$ to $\mathcal{L}(Z|K)$ in the limit $\mathcal{L}(Z,K) = \lim_n \mathcal{L}(Z_n, K_n)$ can be achieved by appropriate discretization of K_n, but the matter is complex. Thus, in principle, the problem is solvable but practical use will require abundant care.

6.9 Possible Extensions

In the previous section we presented a few of the available results for LAN or LAMN families, but we have done so in a restricted environment. That is because simple things are best when first seen in a simple form, not because they are not available in a more complex setting. Among the assumptions made were the following:

(i) $\Theta = \Re^k$, or a subset of $\Re^k$.

(ii) The scaling coefficients δ_n are numbers, not matrices.

(iii) These δ_n do not depend on θ.

(iv) The terms "linear" and "quadratic" refer to the vector structure of $\Re^k$.

(v) The points of interest (for instance, where LAQ holds) are interior points of Θ in $\Re^k$.

If one grants the other assumptions, item (v) above certainly needs modifications for practical use. It would be unfortunate, for instance, to rule out cases where Θ is the positive quadrant of $\Re^2$ and θ is either a boundary point or even the apex of that quadrant.

Such cases can be studied without much difficulty. One needs, of course, to modify appropriately the statements on posterior distributions of Section 6.4. For proper handling, one should introduce tangent spaces, as done by Chernoff [1954]. A short mention of such tangent sets will occur in Chapter 7.

The introduction of rescaling through matrices mentioned in item (ii) above introduces some complications, but they are not insurmountable. The fact that we assumed that the rescaling is *independent of* θ is a more serious matter. There is also the possibility of replacing $\Re^k$ by differentiable manifolds. In these situations, one must be careful of the restriction described in item (iv). There have been a few cases in the literature where the LAN type local structure of an experiment was masked by square root transformations or analogous inappropriate selection of local coordinate systems.

An attempt to avoid all such difficulties occurs in Le Cam [1986, chapter 11]. There, Θ is just a set without any structure of its own. One introduces a vector space in which "linear" and "quadratic" make sense by considering the space $\mathcal{M}_0$ of finite signed measures μ that are carried by finite subsets of Θ. If any vector space can supply a framework for LAN assumptions, that space $\mathcal{M}_0$ will, as could be gathered from the description given in Chapter 4 in the case of gaussian experiments. In addition, one introduces on Θ a substitute for a distance in the form of a function q_n defined by $q_n^2(s,t) = -8 \log \int \sqrt{dP_{s,n} dP_{t,n}}$.

In this manner, one can avoid all forms of rescaling and, for instance, treat cases where $\delta_n = \delta_n(\theta)$ varies arbitrarily as a function of θ. One can even study whether local approximations can be carried out with finite-dimensional vector spaces. The formulation of Le Cam [1986, chapter 11], is a bit involved, but generally not difficult. It was meant to cover many different possibilities. Yet it leaves out what are perhaps the most interesting ones; it is essentially limited to cases that would be deemed "parametric" at least in the limit. (It does cover some "nonparametric" ones as seen in mixtures of Poisson distributions or birth and death processes; see Le Cam and Yang [1988, p. 163–175].

The troubles can be seen to occur in the contiguity assumption made for the LAQ conditions and in the *existence* of estimates that satisfy Assumption 2, Section 6.3. In the general formulation of Le Cam [1986, chapter 11], the auxiliary estimates θ_n^* used here should be such that $q_n(\theta_n{}^*, \theta)$ stays bounded in $P_{\theta,n}$ probability. In most truly nonparametric cases, such estimates do not exist. One could conceive of bypassing the difficulties by using the "method of sieves" of Grenander [1981] and approximating the experiments under study by other experiments, say, $\mathcal{F}_n = \{Q_{\theta,n}\}$, for which finite-dimensional approximations exist. The $\mathcal{F}_n$ might not include any measure $Q_{\theta,n}$ close enough to the "true" $P_{\theta,n}$, thus causing difficulties.

The possibilities do not seem to have been sufficiently studied.

6.10 Historical Remarks

The name LAQ (locally asymptotically quadratic) is taken from a preprint of Jeganathan [1988]. The LAN conditions were introduced by Le Cam [1960]. The more general LAMN conditions were later introduced almost simultaneously by several authors including Davies [1985] (note the date; However, a preprint of the paper was available as early as 1980), Jeganathan [1980], and Swensen [1980]. Numerous other authors were involved, including Basawa and Prakasa Rao [1980], Basawa and Scott [1983], Feigin [1985] and many more.

The method of construction described in Section 6.3 can be found in Le Cam [1974]. A previous description by Le Cam [1960] seemed to depend on unexpected algebraic properties of certain functions. The asymptotic sufficiency properties, for the LAN case, are in Le Cam [1960, 1969, 1974] as well as in [1986]. Similar constructions and properties for the LAMN case are given in Jeganathan [1980]. The local Bayes properties of Section 6.4 go back at least to Laplace [1810a, 1810b, 1820]. They were used in particular cases by Le Cam [1953, 1956]. Le Cam [1986, chapter 12], attempts to give necessary and sufficient conditions for the existence of approximately normal posterior distributions.

The subject is related to what is called the Bernstein–von Mises theorem. That is, However, a different result that says that for *fixed* prior distributions the posterior distributions can be approximately normal, under sufficient regularity and consistency conditions. The remark, in Section 6.5, about the fact that certain invariance and regularity properties hold almost everywhere occurs in Le Cam [1974]. It was further developed by Jeganathan [1982, 1983], who is responsible for showing that the LAQ conditions imply LAMN almost everywhere. He also derives other results of the same character. The fact that approximations of log likelihood by $t'S_n - A(\theta, t)$, where A is nonrandom, already implies LAN almost everywhere is in Le Cam [1960].

As already mentioned, Section 6.6 draws on many sources. The fact that several apparently different conditions for LAMN are equivalent was observed independently by Davies, Jeganathan, Swensen, and perhaps others around 1980.

Anderson's result about convex sets and functions is proved in Anderson [1955]. It gives a neat solution to the problem of evaluating posterior risks. Theorem 1 (the local asymptotic minimax theorem) of Section 6.6 has a long history. A form of it, for the i.i.d. LAN case, occurs in Le Cam [1953] together with an asymptotic admissibility result. There were previous claims, especially by Fisher [1922, 1925] but Fisher's formulation, which did not involve the "$\sup_\tau$" of the theorem and in fact did not say anything about alternatives $\theta + \delta_n \tau$, could not be correct. That a better formulation was needed was shown by Hodges, [1952] (unpublished, but cited in Le Cam [1953]). One of us was told that there had been a previous example by Ville. Similar results, but with restrictions on the form of the estimates, had been given in the i.i.d. case by Edgeworth [1908] and Wilks [1938].

The formulation used here was given in the LAN case by Hájek [1972]. Jeganathan [1980] covered the LAMN case. Hájek's paper [1972] gives a weaker result in its formal statement. However, contrary to claims by Fabian and Hannan [1982], Hájek also gave in the LAN case a result equivalent to the one stated here (see Hájek [1972, p. 189]). Hájek's [1972] paper also contains, for one-dimensional parameters, an asymptotic admissibility statement. Both results were later shown to be consequences of general results of Le Cam [1979]. That the general results of Le Cam [1979] imply Hájek's asymptotic minimax theorem is easy to see. The asymptotic admissibility result is deeper. Le Cam [1974] gives a formulation that covers two-dimensional parameters for usual risk functions. The asymptotic uniqueness theorem of Hájek [1972] has also been given a general form in Le Cam [1979]. That form is curious in that it applies only to nonrandomized procedures that are uniquely admissible for the limit experiments.

The asymptotic minimax results were given much attention for infinite-dimensional parameter sets, by Moussatat [1976] and Millar [1983].

The convolution theorem stated here as Theorem 3 is from Hájek [1970]. Some credit should also be given to Inagaki, who considered a special case. Hájek's proof was complex. It was soon replaced by a simpler argument of Bickel, reproduced in Roussas [1972]. Le Cam [1972] (see also [1986]) gives a formulation valid for certain groups with a proof that does not rely on special properties of the normal distributions except that they form a shift family, dominated by the Lebesgue measure. The proof given here is taken from van der Vaart [1988]. Formulations and proofs of conditional convolution theorems for the LAMN case are in Jeganathan [1983]. Formulations and proofs for the infinite-dimensional LAN case are in Moussatat [1976] and Millar [1985]. The history of the subject is rather complex. One can trace some of it to the thesis of C. Boll [1955]. Neither Boll nor Le Cam [1972, 1986] seem to have been aware of the existence of a paper by J. G. Wendel [1952]. Le Cam [1994] made an attempt to generalize the convolution theorem to a form that would cover van der Vaart's and would not contain restrictions to gaussian distributions or similar. The statement on page 406 of Le Cam [1994] seems to suffer from "excessive youthful enthusiasm." The sketch of the proof given there (p. 407) was never completed. Recently (January 2000), W. Schachermayer and H. Strasser gave a counterexample; additional conditions seem necessary.

Note that Wendel's result [1952] was about the algebra of

finite measures dominated by the Haar measure and linear maps of that space into itself. This structure is retained in Hájek [1970] and Le Cam [1986]. By contrast, van der Vaart [1991] clearly uses two vector spaces, say, $\mathcal{X}$ and $\mathcal{Y}$, and a linear map A from $\mathcal{X}$ to $\mathcal{Y}$. The convolution takes place on $\mathcal{Y}$.

The fact that if a shift experiment on $\mathcal{Y}$ is weaker than a shift experiment on $\mathcal{X}$, it is already weaker than the image of $\mathcal{E}$ by A is true under van der Vaart's conditions, but false in general. We are uncertain about who first noticed that possibility. It is important for semiparametric studies.

The ideas expounded here owe a lot to Wald [1943]. Wald considers only the i.i.d. case under fairly strong regularity conditions. He concentrates his attention on maximum likelihood estimates $\hat{\theta}_n$ and proceeds to claim that one can approximate the experiment $\{P_{\theta,n}; \theta \in \Theta\}$ by another $\{G_{\theta,n}; \theta \in \Theta\}$ where $G_{\theta,n}$ is the gaussian approximation to the distribution $\mathcal{L}[\hat{\theta}_n|\theta]$ of the maximum likelihood estimate. The approximation formulation used by Wald involves set transformations instead of the Markov kernels used later by Le Cam [1960, 1964]. Wald's proof is complex; his conditions are too strong for comfort. A different version, not restricted to maximum likelihood estimates, was later given by Le Cam [1956] under weaker conditions that, however, imply only uniformity on compacts of Θ instead of Θ as a whole. Le Cam [1956] was written to answer a question of J. Neyman about his $C(\alpha)$ tests.

Wald's paper of 1943 goes on to study the implications of the replacement of the experiment $\{P_{\theta,n}; \theta \in \Theta\}$ by the heteroschedastic gaussian $\{G_{\theta,n}; \theta \in \Theta\}$. Here we have mentioned only asymptotic optimality properties for χ^2-like tests or confidence sets based on them. There is a lot more in Wald's paper; some of it is reproduced in Le Cam [1986, chapter 11, section 9].

That precautions must be taken with Wald's formulation (and therefore also for the formulation used here or in Le Cam [1986]) was pointed out by Hauck and Donner [1977] and especially by Vaeth [1985]. An attempt to describe how to take precautions was given by J. S. Hodges paper [1987].

We have said little about infinite-dimensional parameter spaces Θ instead of finite-dimensional, even though there are extensive results for that case. One thing that should be emphasized is that, in infinite dimensions, there is no good equivalent of the Lebesgue measure, so all the "almost everywhere Lebesgue" statements fall apart. It may be possible to use relations involving Kolmogorov's metric entropy. For other approaches, see, for instance, Millar [1983, 1985] and Donoho and Liu [1991]. Our reason for not presenting a study of infinite-dimensional cases is simple: The construction of Section 6.3 usually cannot be carried out, because the appropriate auxiliary estimates do not exist. (By that we mean that it is a theorem that they do not exist.) One needs properties in addition to the very strictly local LAN or LAMN.

7

Independent, Identically Distributed Observations

7.1 Introduction

The structure that has received by far the most attention in the statistical literature is one that can be described as follows. One takes a family of probability measures $\{p_\theta : \theta \in \Theta\}$ on some space $(\mathcal{X}, \mathcal{A})$. Then one considers experiments $\mathcal{E}_n = \{P_{\theta,n}; \theta \in \Theta\}$ where $P_{\theta,n}$ is the joint distribution of n observations $X_1, X_2, \ldots, X_n$ all independent with individual distribution p_θ. One studies the asymptotic behavior of the system as n tends to infinity. We shall refer to this as the "standard i.i.d. case."

Here we shall use such systems to illustrate the workings of the LAN conditions of Chapter 6.

To illustrate the LAMN conditions we shall let the number N of observations $X_1, X_2, \ldots, X_N$ be a random variable, stopping time of the infinite sequence $X_1, X_2, \ldots, X_n, \ldots$.

This is just meant as an illustration, not as a recommendation for practical use. It must be pointed out that the asymptotics of the "standard i.i.d. case" are of little relevance to practical use of statistics, in spite of their widespread study and use. The reason for this is very simple: One hardly ever encounters fixed families $\{p_\theta : \theta \in \Theta\}$ with a number of observations that will tend to infinity. There are not that many particles in the visible universe! The use of such considerations is an abuse of confidence that has been foisted upon unsuspecting students and practitioners owing to the fact that we, as a group, possess limited analytical abilities and, perforce, have to limit ourselves to simple problems. This is not heresy. It has been said by many; see, for instance, Lehmann [1949].

However, it has not been said enough. Common textbooks often include statements about "consistency" of estimates. Apparently it is considered good for estimates $\hat{\theta}_n$ to converge in probability to the "true value" θ as $n \to \infty$. It is even better if they do so almost surely. Few texts point out that such a property is entirely irrelevant to anything of value in practice. There one may wish to find estimates $\hat{\theta}_n$ that take values close to a θ such that p_θ is a fairly accurate description of the distribution of the underlying variables or, failing the availability of an accurate description in the family $\{p_\theta : \theta \in \Theta\}$, as close as can be within the limitations of that family. If you have, say, 10^6 observations, it is of acute irrelevance what $\hat{\theta}_n$ does along the sequence 10^{100k}, $k = 1, 2, \ldots$, or along the sequence of prime numbers larger than 2^{10231}.

The arguments are not even in accord with standard practice. Almost any statistician will use families $\{p_\theta : \theta \in \Theta_n\}$ where the "number of parameters" depends on the "number n of observations."

That is usually done by considering more complex models when the available information becomes better and more complete. It has some relation to the "number of parameters" but the relation is not clearcut.

The use of asymptotics "as $n \to \infty$" for the standard i.i.d. case seems to be based on an entirely unwarranted act of faith. If you do prove that $\hat{\theta}_n$ has some good asymptotic property, maybe some of that will percolate down to $n = 10^6$ or even $n = 100$. You might even be able to check it occasionally with some Monte Carlo experiment.

One obtains a better view of what can happen if one goes slightly off the standard i.i.d. formulation by taking n independent observations from measures $\{p_{\theta,n}; \theta \in \Theta_n\}$ where $p_{\theta,n}$ or Θ_n or both depend on n. That gives a context where "consistency" means little, where the difference between "fixed alternatives" and "near alternatives" has lost its meaning and where limit theorems "as n tends to infinity" can take many different shapes. Consider, for instance, a rare disease that occurs in a human population of size n with a frequency p_θ. We all know that the human population is, unfortunately, "tending to infinity" but what should the asymptotics be if, in all history, only two cases of the disease have even been mentioned or observed anywhere?

However, we digress and will still use the standard i.i.d. case as an illustration.

7.2 The Standard i.i.d. Case: Differentiability in Quadratic Mean

Let $\{p_\theta; \theta \in \Theta\}$ be a family of probability measures on a space $(\mathcal{X}, \mathcal{A})$. Let $P_{\theta,n}$ be the joint distribution of n independent observations, each distributed according to p_θ.

In this situation, when Θ is a subset of a euclidean space, it often happens that a certain condition of differentiability in quadratic mean is satisfied. It is sufficient to imply that the $P_{\theta,n}$ will satisfy the LAN conditions with scaling constants $\delta_n = n^{-1/2}$. The condition can be stated as follows. Let θ be a particular point of Θ. For another point $\tau \in \Theta$, let $Z(\tau) = \sqrt{dp_\tau/dp_\theta}$ considered as a random variable for the distribution induced by p_θ. As before dp_τ/dp_θ is the density of the part of p_τ that is dominated by p_θ.

(DQM_o) *There are random vectors V_θ, with $E_\theta\|V_\theta\|^2 < \infty$, such that*

$$\lim_{\tau\to\theta} E_\theta\{\frac{1}{|\tau-\theta|}[Z(\tau) - Z(\theta) - \frac{1}{2}(\tau-\theta)'V_\theta]\}^2 = 0.$$

Furthermore, let $\beta(\tau,\theta)$ be the mass of the part of p_τ that is p_θ singular, then $\beta(\tau,\theta)/|\tau-\theta|^2 \to 0$ as $\tau \to \theta$.

There are variations on that condition. It is often neater to ask for differentiability in quadratic mean at θ of the process whose covariance kernel is $\int \sqrt{dp_s dp_t}$. Then the extra condition concerning singular parts is automatically satisfied if θ is interior to Θ or, more generally, if the contingent of Θ at θ is sufficiently rich; see, for instance, Le Cam [1986, p. 575]. The contingent of Θ at θ is the set of rays $\{\alpha u; \alpha \geq 0\}$ where u is a cluster point of a sequence $u_n = (\tau_n - \theta)/|\tau_n - \theta|$, with $\tau_n \in \Theta$ tending to θ.

It is clear that definitions such as (DQM_o) will not give much information for vectors v that lie outside the contingent.

There are several other sets of conditions that have been (and are) used in the literature. One such set can be found in Cramér's [1946] famous book. Cramér assumed that the p_τ; $\tau \in \Theta$ had densities with respect to some dominating measure and that suitably selected versions $f(x,\tau)$ of these densities were three times differentiable in τ for each fixed x. He assumed in addition certain conditions of "differentiability under the integral sign."

Nowadays conditions similar to Cramér's are often stated in terms of differentiability of $\log f(x,\tau)$. A possible set is as follows:

(C1) The p_τ are mutually absolutely continuous.

(C2) There are vector-valued functions $(x,\theta) \rightsquigarrow \varphi(x,\theta)$ and matrix-valued functions $(x,\tau) \rightsquigarrow B[x,\tau]$ with $B[x,t]$ defined at least on segments $t = (1-\xi)\theta + \xi\tau$; $\xi \in [0,1]$, for τ in a neighborhood of θ relatively to Θ.

(C3) In such a neighborhood, $\log f(x,\tau)$ admits a Taylor expansion

$$\begin{aligned} \log \frac{f(x,\tau)}{f(x,\theta)} &= (\tau-\theta)'\varphi(x,\theta) \\ &\quad -(\tau-\theta)' \int_0^1 (1-\xi)B[x,t(\xi)](\tau-\theta)d\xi \end{aligned}$$

with $t(\xi) = (1-\xi)\theta + \xi\tau$.

(C4) $E_\theta\varphi(X,\theta) = 0$ and $E_\theta[\varphi(X,\theta)\varphi'(X,\theta)] = E_\theta B(X,\theta)$.

(C5) θ admits (relatively to Θ) a compact neighborhood in which the trajectories $\tau \rightsquigarrow B(x,\tau)$ are continuous and bounded componentwise by some function $H(x)$ such that $E_\theta H(X) < \infty$.

It can be shown that such conditions imply our DQM_o condition. An indirect proof is given in Le Cam [1986, p. 584]. Cramér's conditions are often satisfied. They imply in neighborhoods of θ a behavior of $\Lambda_n(\tau,\theta) = \log(dP_{\tau,n}/dP_{\theta,n})$ that is much nicer than that implied by the DQM_o. They imply that the log likelihoods can be approximated by quadratic expressions *uniformly* on small neighborhoods in the sense that

$$\sup_{\{\tau:|\tau-\theta|<\epsilon_n\}} |\Lambda_n(\tau,\theta) - (\tau-\theta)'\sum_{j=1}^{n}\varphi(X_j,\theta) + \frac{1}{2}(\tau-\theta)'\sum_{j=1}^{n}B(X_j,\theta)(\tau-\theta)|$$

will tend to zero in probability if ϵ_n does. This is indeed a lot more than what was used in the LAN assumptions. These would involve only sets of the form $\{\tau : |\tau-\theta| \leq b/\sqrt{n}\}$ and would not support a supremum taken over τ in such sets.

Let us pass to some implications of DQM_o. To express them in simple notation *we shall assume that* $\theta = 0$. We shall let $h^2(\tau)$ be the square Hellinger distance $(1/2)\int(\sqrt{dp_\tau} - \sqrt{dp_\theta})^2$ and denote simply by $\frac{V}{2}$ the derivative in quadratic mean of $Z(\tau)$ at $\theta = 0$.

With these conventions and notation we have the following relations.

Lemma 1. *Let* $\tau_m \in \Theta$ *be a sequence such that* $\tau_m \to \theta = 0$ *and such that* $u_m = \tau_m/|\tau_m| \to u$. *Then* $E\{[Z(\tau_m) - 1]/|\tau_m|\}^2$ *and* $2h^2(\tau_m)/|\tau_m|^2$ *tend to* $E(u'V)^2/4$, *where* E *denotes expectation under* p_0.

Proof. According to Section 5.3, one has $E[Z(\tau_m) - 1]^2 = 2h^2(\tau_m) - \beta(\tau_m,\theta)$ where $\beta(\tau_m,\theta)$ is the mass of the part of p_τ that is p_θ-singular. By the second part of DQM_o these singular parts may be neglected.

The convergence in quadratic mean included in DQM_o implies that

$$E[\frac{1}{|\tau_m|}(Z(\tau_m) - 1)]^2 \to E(\frac{u'V}{2})^2.$$

Hence the result. □

Lemma 2. *If* DQM_o *holds at* $\theta = 0$ *then* $Eu'V = 0$ *for every* u *in the contingent of* Θ *at* $\theta = 0$.

Proof. Convergence in quadratic mean under p_θ implies convergence in the first mean. Thus

$$E\frac{1}{|\tau_m|}[Z(\tau_m)-1] = -\frac{h^2(\tau_m)}{|\tau_m|} \to \frac{1}{2}E(u'V).$$

Since $h^2(\tau_m)/|\tau_m|^2$ tends to a finite limit this implies that $E(u'V)=0$. □

To state our next proposition, recall that $P_{\theta,n}$ is the distribution of the sequence $\{X_1, X_2, \ldots, X_j, \ldots, X_n\}$. Every random variable defined on $\{\mathcal{X}, \mathcal{A}, p_\theta\}$ admits a corresponding copy defined in function of the jth variable X_j. This will be indicated below by adding a subscript j. Thus, for instance, $\frac{V_j}{2}$ is the derivative in quadratic mean for measures $p_{\tau,j}$ and $p_{\theta,j}$ defined on the jth replicate $(\mathcal{X}_j, \mathcal{A}_j)$ of $(\mathcal{X}, \mathcal{A})$. This gives the following result.

Proposition 1. *Let DQM_o be satisfied at θ. Then, for any bounded sequence $\{t_n\}$ such that $\theta + n^{-1/2}t_n \in \Theta$ the difference*

$$\Lambda_n(\theta + \frac{1}{\sqrt{n}}t_n;\theta) - \{\frac{1}{\sqrt{n}}t'_n(\sum_{j=1}^n V_j) - \frac{1}{2}t'_n(EVV')t_n\}$$

tends to zero in $P_{\theta,n}$ probability as $n\to\infty$.

Furthermore, for each u in the contingent of Θ at θ the sum $(1/\sqrt{n})u'\sum_{j=1}^n V_j$ is asymptotically centered normal with variance
$E(u'V)^2$.

Proof. Assume $\theta = 0$. According to the preceding lemmas the DQM_o condition may be written in the form:

$$Z(\tau_n) - 1 = \frac{1}{2}\tau'_n V - h^2(\tau_n) + |\tau_n|R(\tau_n)$$

where $EV = 0$, $ER(\tau_n) = 0$ and $ER^2(\tau_n)$ tends to zero as $\tau_n \to \theta = 0$. Applying this to replicates of these variables with $\tau_n = \theta + n^{-1/2}t_n$ yields

$$\sum_{j=1}^n [Z_j(\theta + \frac{t_n}{\sqrt{n}}) - 1] \quad = \quad \frac{1}{2}\frac{1}{\sqrt{n}}\sum_{j=1}^n t'_n V_j - nh^2(\theta + \frac{t_n}{\sqrt{n}})$$

$$+|t_n|\frac{1}{\sqrt{n}}\sum_{j=1}^{n} R_j(\theta + \frac{t_n}{\sqrt{n}}).$$

The third term on the right has expectation zero and a variance that tends to zero. Thus it tends to zero in quadratic mean. The second term is, by Lemma 1, equivalent to

$$\frac{1}{8}|t_n|^2 E[\frac{t_n'}{|t_n|}V]^2 = \frac{1}{8}E(t_n'V)^2.$$

By Lemma 2 and the Central Limit Theorem $n^{-1/2}\sum_{j=1}^{n} t_n' V_j$ is asymptotically normal with mean zero and variance $E(t_n'V)^2$.

Now use the Taylor expansion $\log(1+x) \sim x - (1/2)x^2$ or Lemma 3, Section 5.3, to pass from the sum of square roots to the sum of logarithms. This will give the desired answer. □

The reader will note that Proposition 1 gives a quadratic approximation to the logarithm of likelihood ratios of the kind we described under the name of LAQ, LAMN, or LAN conditions in Chapter 6. Thus we have the following:

If θ is an interior point of Θ, all the results of Chapter 6 are applicable here.

In particular, for locally concentrating prior distributions, the posterior distributions will be asymptotically normal. In addition, the Hájek–Le Cam asymptotic minimax theorem will be applicable. If there exist appropriately behaved auxiliary estimates θ_n^*, the construction of Section 6.3 will give centers Z_n that are asymptotically normally distributed, asymptotically "efficient," sufficient etc.

In Chapter 6 we assumed, for simplicity, that $\Theta = \Re^k$ or at least that the true θ was interior to Θ in $\Re^k$. This has not been assumed for the derivation of Proposition 1; the condition DQM_o involves only neighborhoods of θ in Θ. If these neighborhoods are rich enough, standard asymptotic optimality properties can easily be derived. If, on the contrary, these neighborhoods are thin or limited, problems may arise.

The words "thin" or "rich" are used here in a nonspecific way. One can illustrate them by example. For this, we shall use subsets Θ of $\Re^2$ with $\theta_0 = 0 \in \Theta$.

For a first example, one can take the set Θ_r of points whose coordinates are rational numbers. It may look sparse, but for most purposes it is rich enough, as we shall see when discussing tangents in the Chernoff sense later in this section.

Another example of a set rich in some directions but not in others would be the set Θ_q formed by the quadrant of points whose coordinates are nonnegative.

For a totally different kind of behavior let us look at a set Θ_s defined in polar coordinates (α, ρ) for an angle α and distance to the origin ρ. Let Θ_s consist of $\theta_0 = 0$ and of the points $(\alpha, \rho(\alpha))$ where $\rho(\alpha) = \exp[-\exp \alpha]$, $\alpha \in (1, \infty)$. We shall call it the "fast spiral."

In the case of Θ_r or Θ_q there is no difficulty in carrying out the construction of centers Z_n described in Section 6.3, provided suitable auxiliary estimates θ_n^* are available. There we used fixed bases $(u_1, u_2, \ldots, u_k)$, but this is unnecessary.

Assume that the derivative in quadratic mean $V/2$ is such that EVV' is nonsingular. One can take bases $(u_1, u_2, \ldots u_k)_n$ that depend on n in a totally arbitrary way subject to the following requirements:

(i) $\theta_n^* + n^{-1/2} u_i \in \Theta$,

(ii) the bases $(u_1, \ldots, u_k)_n$ stay bounded, and

(iii) the determinant of $(u_1, \ldots u_k)_n$ does not tend to zero.

(A proof using such arbitrary bases is given in Le Cam [1986, chapter 11].

It is not difficult to find such bases for Θ_r or Θ_q; for the fast spiral Θ_s it is quite impossible. The part of the spiral $\{(\alpha, \rho(\alpha))\colon \alpha \geq \alpha_1\}$ for α_1 large, looks to all intents and purposes like a segment of a straight line going from the origin to the point $(\alpha_1, \rho(\alpha_1))$.

If the actual distribution of the n observations arises from $\theta_0 = 0$, the only relevant part of the spiral Θ_s will be the part at distance $b/\sqrt{n}$ of zero, with b equal to 10 or perhaps 20. At that magnification it will look like a line segment (whose direction varies slowly as n changes). It will not support any genuine two-dimensional asymptotics.

For the quadrant Θ_q one must note that the centers Z_n of Section 6.3, will often lie out of Θ_q. This means that they are not very good as *estimates* for θ's near $\theta_0 = 0$. However, projecting them on Θ_q would make them lose their asymptotic sufficiency properties.

The two sets Θ_r and Θ_q behave quite differently if one is interested in posterior distributions. For Θ_q the arguments of Section 6.4 remain applicable provided one restricts all measures to Θ_q and renormalizes them. For Θ_r the situation is more complex since Θ_r does not support any continuous prior measures. The arguments of Section 6.4 could still be applied with appropriate modifications that need complicated notation.

As far as asymptotic minimax risks and convolution theorems are concerned the set Θ_r behaves essentially in the same way as $\Re^2$ itself. For Θ_q the situation is more complex near zero, but the asymptotic minimax risk stays the same; see Proposition 2.

In Lemma 2, we have used the *contingent* of Θ at θ. This is always a closed cone and thus relatively nice. It need not give a good picture of what Θ looks like near θ. Take, for instance our fast spiral Θ_s. The contingent of Θ_s at zero is the entire space $\Re^2$. However, if one takes a finite set, say, $\{v_1, v_2\}$ in the contingent, there is no assurance that for a given n, however large, one can find points of the type $\theta + n^{-1/2}t_{i,n} \in \Theta$, $i = 1, 2$ such that $t_{i,n}$ is close to v_i and the $t_{i,n}$ stay bounded away from zero and infinity.

For instance let v_1 be the point $(1, 0)$. To get a point on the spiral such that $t_{1,n}$ is close to $(1, 0)$ one must take an angle α close to $2\pi\nu$ for some integer ν. For instance, take $\alpha_0 = 2\pi \times 2$; then a variation on α of the order of $1/1000$ will change $\rho(\alpha)$ by a factor of about 5×10^{123}. The next time the spiral passes close to a multiple of $(1, 0)$ the value of ρ will be utterly negligible compared to that at $\alpha_0 = 4\pi$. For any given value of n, with n large, the part of the scaled-up set $\sqrt{n}\Theta$ that is at reasonable distance, say, within 10, of zero will look more like a segment of line with a little blob at zero.

Another set that has been used is the "tangent space" at zero. To distinguish it from the contingent, let us call it the "continuous path tangent space" or CP Tangent. It is a cone defined by limits $\lim_{\epsilon \to 0, \epsilon > 0} \varphi(\epsilon)/\epsilon$ where φ is a continuous function from $[0, 1]$ to Θ with $\varphi(0) = 0$. The only φ's to be used are those for which the limit exists. In the spiral example there are no such functions except the function $\varphi(\epsilon) \equiv 0$.

Another concept of "tangency" was introduced in statistics by Chernoff [1954]. A set T is said to be tangent to Θ at $\theta = 0$ in the sense of Chernoff if, denoting by $B(\epsilon)$ the ball centered at zero that has radius ϵ, the Hausdorff distance between $T \cap B(\epsilon)$ and $\Theta \cap B(\epsilon)$ tends to zero faster than ϵ as ϵ tends to zero. The Hausdorff distance between two sets A and B is the maximum of the two numbers

$$\sup_t \inf_s \{d(s,t); s \in A, t \in B\}$$

and

$$\sup_s \inf_t \{d(s,t); s \in A, t \in B\},$$

where $d(s, t)$ is the distance between the points s and t.

In a case such as our spiral this will not work well either, if one keeps the set T fixed. However, for each n large, one can replace the part Θ_n of Θ that is most interesting by another set T_n such that the Hausdorff distance between $\sqrt{n}T_n$ and $\sqrt{n}\Theta_n$ is small.

An experiment $\{P_{\theta,n} : \theta \in \Theta_n\}$ can be reindexed by T_n. For each $t \in T_n$, let $\theta(t)$ be chosen so that $|\theta(t) - t| = n^{-2} + \inf_\theta\{|\theta - t|;\ \theta \in \Theta_n\}$. This will give a family $\{P_{\theta(t),n}; t \in T_n\}$. This technique is useful very generally to replace curves or surfaces by straight lines, planes or hyperplanes. See, for instance, Neyman's theory of BAN estimates (Neyman [1949]) or arguments replacing manifolds by tangent spaces (see Le Cam [1986, chapter 11, section 7]. Here it would permit the replacement of the rational space Θ_r by $\Re^2$.

For any continuous bounded loss function of the form $W[\sqrt{n}\,(Z_n - \theta)]$, replacing Θ_n by a tangent T_n in the sense of Chernoff is not going to matter asymptotically. In addition, a difference in the expectations of normal distributions that tends to zero faster than $1/\sqrt{n}$ will not matter asymptotically.

To indicate some other possibilities consider a set $\Theta \subset \Re^k$ with $\theta = 0 \in \Theta$. Let us say that a set $T \subset \Re^k$ is a subtangent of Θ at $\theta = 0$ if for every finite subset $\{v_1, v_2, \ldots, v_m\} \subset T$ and every sufficiently large n, there are points $t_{1,n}, \ldots, t_{m,n}$ such that $\theta + n^{-1/2} t_{i,n} \in \Theta$ and $t_{i,n} \to v_i$.

One can then state a result of the following nature.

Proposition 2. *Assume that T is a convex cone with an interior point. Assume that $\mathcal{E} = \{p_\theta; \theta \in \Theta\}$ satisfies DQM_o at $\theta = 0$. Assume also that T is a subtangent of Θ at $\theta = 0$. Then for any bowl shape loss function the asymptotic minimax risk is not smaller than if θ was an interior point of $\Re^k$.*

We shall not give the proof here. It can be obtained from a more general result of van der Vaart [1988]. However, note that for each sufficiently large n and each finite set $\{v_1, \ldots, v_m\}$, one can associate points $t_{i,n}$ such that $\theta + n^{-1/2} t_{i,n} \in \Theta$ and $t_{i,n}$ is close to v_i. Doing this in a systematic way will yield experiments $\{P_{\theta + n^{-1/2} t_{i,n}, n}; i = 1, \ldots, m\}$ that tend weakly to the gaussian shift limit experiment restricted to the same set $\{v_1, v_2, \ldots, v_m\}$. Thus it is enough to evaluate the minimax risk of that gaussian experiment. Since a convex cone with an interior point contains balls as large as one pleases, the minimax value cannot be any smaller than it is in such a ball.

To terminate this section, let us note that Proposition 1 admits a converse.

Proposition 3. *Let $\mathcal{E} = \{p_\theta; \theta \in \Theta\}$ be such that for any bounded sequence $\{t_n\}$ such that $\theta + n^{-1/2} t_n \in \Theta$ the difference*

$$\Lambda_n(\theta + \frac{1}{\sqrt{n}} t_n; \theta) - \frac{1}{\sqrt{n}} t_n' (\sum_{j=1}^{n} X_j) + \frac{1}{2} t_n' (EXX') t_n$$

tends to zero in $P_{\theta,n}$ probability for variables X_j that are copies on $(\mathcal{X}_j, \mathcal{A}_j)$ of a variable X on $(\mathcal{X}, \mathcal{A})$ such that $E_\theta X = 0$ and $EX'X < \infty$. Then $\mathcal{E}$ satisfies (DQM_o) at θ.

For a proof, see, for instance, Le Cam [1986, p. 584].

In spite of this, one can construct examples of experiments $\mathcal{E} = \{p_\theta; \theta \in \Re\}$ such that the local experiments $\{P_{\theta+n^{-1/2}t,n}; t \in \Re\}$ with $\theta = 0$ are asymptotically gaussian shift, $\mathcal{G} = \{Q_t,\ t \in R\}$ with $Q_t = \mathcal{N}(t,1)$, and p_τ admits a density $(dp_\tau/d\lambda)(x) = f(x,\tau)$ with respect to the Lebesgue measure but $\lim_{\tau\to 0} f(x,\tau) / \tau^m = 0$ almost surely for all integers m; see Le Cam [1986, p. 583].

7.3 Some Examples

In light of the results of Chapter 6 and the conditions of differentiability in quadratic mean, we shall give some examples of families of densities that have been encountered in the literature.

Example 1. The Cauchy density $f(x,\theta) = 1/\pi(1 + (x - \theta)^2)$ with respect to the Lebesgue measure on $(-\infty, +\infty)$.

We say that DQM_o is satisfied at all $\theta \in \Re$. This can be verified directly without trouble. However, it is also a particular case of an observation of Hájek [1962], who proved the following:

Lemma 3. *Let f be a probability density with respect to the Lebesgue measure λ. Assume that (1) f is absolutely continuous and (2) the Fisher information $\int[(f')^2/f]d\lambda$ is finite. Then the shift family with densities $f(x-\theta)$, $\theta \in \Re$ satisfies DQM_o at all θ.*

This gives an easy way of checking DQM_o for many shift families. For other families with densities $f(x,\theta)$ with respect to some finite measure μ, one can use an extension of this criterion as follows.

Lemma 4. *Suppose $\theta \rightsquigarrow f(x,\theta)$ is absolutely continuous in $\theta \in R$ for each x. Let $f'(x,\theta) = \frac{\partial}{\partial\theta} f(x,\theta)$ where it exists. Take $f'(x,\theta) = 0$ otherwise. Let $\sigma^2(\theta) = \int[(f'(x,\theta))^2/f(x,\theta)]\mu(dx)$.*

If

$$\frac{1}{2\epsilon}\int_{-\epsilon}^{+\epsilon}|\sigma(\theta+\tau)-\sigma(\theta)|d\tau \to 0$$

as $\epsilon \to 0$, *then* DQM_o *is satisfied at* θ.

For a proof, see Le Cam [1986, p. 588]. It should perhaps be mentioned that the conditions of Lemma 4 are sufficient but not necessary for DQM_o. Differentiability in quadratic mean at a point θ does not imply pointwise differentiability, nor does it imply continuity in θ of trajectories, much less absolute continuity.

It should also be noted that the criterion applies to parameter $\theta \in \Re$, not in $\Re^k$ for $k > 1$. For $\theta \in \Re^k$, $k > 1$, it may be simpler to use convergence in measure of the ratios $[f(x,\tau) - f(x,\theta) - (\tau-\theta)'V]/|\tau-\theta|$ together with appropriate convergence of the square norms $\int[\sqrt{f(x,\tau)} - \sqrt{f(x,\theta)}/|\tau-\theta|]^2\mu(dx)$ to $E(u'V)^2/4$ if $(\tau-\theta)/|\tau-\theta| \to u$. For other possibilities, see Dieudonné [1960, p. 151-152].

The Cauchy distribution can be used to illustrate several other phenomena.

Instead of verifying DQM_o directly, it is often more convenient to use densities $f(x,\theta)$ with respect to a dominating measure μ. Then one evaluates upper bounds for the quantities $h^2(s,t) = (1/2)\int[\sqrt{f(x,s)} - \sqrt{f(x,t)}]^2\mu(dx)$ for s close to t. The condition

$$\limsup_{s\to t}\frac{h(s,t)}{|s-t|} < \infty$$

at almost all t will imply differentiability in quadratic mean at almost all t, thus DQM_o at almost all points t that are interior to Θ in $\Re^k$. For a proof see Le Cam [1970] or [1986].

For a shift family, like the Cauchy, all points are alike. Thus "almost all" becomes "all."

Another remark is that Section 6.3 makes use of auxiliary estimates θ_n^* such that $\delta_n^{-1}|\theta_n^* - \theta|$ stays bounded in probability. Here δ_n^{-1} can be taken equal to $\sqrt{n}$. For a distribution such as the Cauchy that is symmetric with a positive continuous density at its center, the empirical median can play the role of θ_n^*, except that it is not "discretized" enough to apply the results of Chapter 6, that is, Assumption 2 in Section 6.3 is not satisfied.

The discretization is there only to avoid bumps and peculiarities in the log likelihood function. For the Cauchy distribution, Cramér's conditions are satisfied. They imply that

$$\sup_{|t|\leq b} |\Lambda_n(\theta + \frac{t}{\sqrt{n}}; \theta) - [t'S_n - \frac{1}{2}t'\Gamma(\theta)t]|$$

tends to zero in $P_{\theta,n}$ probability.

A look at the arguments of Chapter 6 shows that this is sufficient to avoid the need for "discretization."

Of course Cramér's conditions are far more restrictive than DQM_o. However, *in the case of one-dimensional parameters*, a condition such as DQM_o valid in a neighborhood of a given θ almost implies that discretization is unneeded. This is only an "almost." To be precise a condition is as follows.

Let $\{Z(t); t \in \Re\}$ be a process whose covariance is $EZ(s)Z(t) = \int \sqrt{dp_s dp_t}$. Assume $Z(t) \geq 0$ almost surely for each t and assume that $\limsup_{s \to t} h(s,t)/|s-t| = \sigma(t) < \infty$ for all t in a neighborhood of a point θ.

Then if $\int_{-\epsilon}^{+\epsilon} \sigma(\theta + t)dt < \infty$ for some $\epsilon > 0$ the log likelihood functions $\Lambda_n(\theta + t/\sqrt{n}; \theta)$; $|t| \leq b$ will satisfy equicontinuity properties that will dispense one from the need of discretization of auxiliary estimates (see Le Cam [1970]).

Here again this is valid for the Cauchy distribution and any one-dimensional shift family $\theta \rightsquigarrow f(x-\theta)$ with $x \to f(x)$ absolutely continuous and finite Fisher information. The extension to families that are not shift families is not difficult.

Example 2. This example, and generalizations of it, was treated by Prakasa Rao [1968].

Fix an $\alpha > 0$ and let the function $f_\alpha(x) = c(\alpha)\exp\{-|x|^\alpha\}$ where $c(\alpha) = [2\Gamma(1/\alpha)]^{-1}$. From this, one can obtain a shift family of densities $f_\alpha(x-\theta) = c(\alpha)\exp\{-|x-\theta|^\alpha\}$. Here x and θ are both in $\Re$.

For fixed α, the function $\log f_\alpha(x) = \log c(\alpha) - |x|^\alpha$ admits a derivative $\varphi(x) = -\alpha \text{sign}(\text{x})|\text{x}|^{\alpha-1}$ except at the point $x = 0$.

For $\alpha > 1/2$ the integral $\int [\varphi(x)]^2 f(x)dx$ is finite. Thus, according to Hájek's criterion (Lemma 3) the family $\{p_\theta; \theta \in \Re\}$ of measures with densities $f_\alpha(x-\theta)$ satisfies DQM_o at all $\theta \in \Re$. However, for $\alpha \in (0, 1/2]$ the integral is infinite, and Hájek's criterion is not applicable, but it suggests that DQM_o is not satisfied.

To find out what happens for $\alpha \in (0, 1/2]$, it is preferable to evaluate the Hellinger distance $h(s,t)$ for points s and t that are close to each other. According to Chapter 5, the product measures $P_{s,n}$ and $P_{t,n}$ will not separate entirely provided $nh^2(s,t)$ stays bounded.

The difference $\sqrt{f(x-t)} - \sqrt{f(x-s)}$ is smooth except for two peaks at $x = s$ and $x = t$. By symmetry one can reduce oneself to evaluation of integrals from $(s+t)/2$ to infinity. Assuming $s < t$ and letting $\tau = t - s$, one sees that the square Hellinger distance is

$$h^2(\tau) = c(\alpha) \int_{-\frac{\tau}{2}}^{\infty} \{\exp\{-\frac{1}{2}|x|^\alpha\} - \exp\{-\frac{1}{2}(x+\tau)^\alpha\}\}^2 dx.$$

Fix an $\epsilon > 0$ and assume that $\tau \in (0, \epsilon)$ tends to zero. According to Lemma 3 the ratio

$$\frac{1}{\tau^2} \int_\epsilon^\infty \{\exp[-\frac{1}{2}x^\alpha] - \exp[-\frac{1}{2}(x+\tau)^\alpha]\}^2 dx$$

will tend to the limit

$$(\frac{\alpha}{2})^2 \int_\epsilon^\infty e^{-x^\alpha} x^{2(\alpha-1)} dx < \infty.$$

Thus we need only to consider the behavior of the integrals taken between $-\tau/2$ and $\epsilon > 0$. Let us look first at the integral from zero to ϵ. Expanding the exponentials by Taylor's formula shows that for ϵ small the integral

$$\int_0^\epsilon \{\exp[-\frac{1}{2}x^\alpha] - \exp[-\frac{1}{2}(x+\tau)^\alpha]\}^2 dx$$

will have a first-order term (order in ϵ) equal to

$$\frac{1}{4}\int_0^\epsilon [(x+\tau)^\alpha - x^\alpha]^2 dx = \frac{1}{4}\tau^{1+2\alpha} \int_0^{\epsilon/\tau} [(y+1)^\alpha - y^\alpha]^2 dy.$$

Now for $0 < \alpha < 1/2$, the integral $\int_1^{\epsilon/\tau}[(y+1)^\alpha - y^\alpha]^2 dy$ will tend to a finite limit as $\tau \to 0$. On the contrary if $\alpha = 1/2$ the integral $\int_1^{\epsilon/\tau}[(y+1)^{1/2} - y^{1/2}]^2 dy$ tends to infinity as $\tau \to 0$ but the ratio

$$\frac{1}{|\log \tau|} \int_1^{\epsilon/\tau} [(y+1)^{1/2} - y^{1/2}]^2 dy$$

tends to a finite limit.

To summarize, let $h^2(\tau)$ be the square Hellinger distance between s and $s+\tau$, $\tau > 0$. Then one can state the following.

Lemma 5. *Let h^2 be as defined and let τ tend to zero. Then the following hold:*

(i) If $\alpha > 1/2$, then $h^2(\tau)/\tau^2$ tends to a finite limit and DQM_o holds.

(ii) If $\alpha = 1/2$, then $h^2(\tau)/\tau^2|\log \tau|$ tends to a finite limit as $\tau \to 0$.

(iii) If $\alpha \in (0, 1/2)$, then $h^2(\tau)/\tau^{1+2\alpha}$ tends to a finite limit.

In cases (ii) and (iii) the part of the integrals situated above $\epsilon > 0$ or below $-\epsilon$ will not contribute to the stated limits.

Proof. This does not quite follow from the above because we have ignored the integrals from $-\tau/2$ to 0 and have said little about the integrals from 0 to τ. In case (ii) these pieces are visibly negligible. In case (iii) they do contribute to the first-order term a quantity $(\tau^{1+2\alpha}/4)\int_{-1/2}^{1}[(y+1)^\alpha - |y|^\alpha]^2 dy$ with the corresponding finite nonzero contribution to the limit. The results are then obtainable by first letting τ and then ϵ tend to zero. $\square$

According to the discussion leading to Assumption $(A'1)$ of Chapter 5, the product measures $P_{\theta,n}$ and $P_{\theta+\tau_n,n}$ will not separate entirely if $nh^2(\tau_n)$ stays bounded.

From this and Lemma 5 one concludes easily that for n independent observations the increments in the values of the parameter that will lead to nondegenerate and not "perfect" behavior of the experiments will be respectively as follows:

(i) for $\alpha > 1/2$, of order $1/\sqrt{n}$.

(ii) for $\alpha = 1/2$, of order $1/\sqrt{n \log n}$.

(iii) for $\alpha \in (0, 1/2)$, of order $1/n^{\beta}$, $\beta = 1/(1 + 2\alpha)$.

This suggests looking at the behavior of the process $\Lambda_n[\theta + \delta_n t; \theta]$ for values of δ_n as specified above. Of course, we already know that for $\alpha > 1/2$ the condition DQM_o holds with the attending behavior. For $\alpha \leq 1/2$ the condition DQM_o does not hold.

Lemma 6. *For $\alpha \in (0, 1/2]$ the process $t \rightsquigarrow \Lambda_n[\theta + \delta_n t; \theta]$ has a limit that is a gaussian process, say, $t \rightsquigarrow Z(t)$. Let $Z_0(t) = Z(t) - EZ(t)$. Then the following hold:*
(i) if $\alpha = 1/2$, $Z_0(t) = tW$ where W is a centered gaussian variable, and
(ii) if $\alpha \in (0, 1/2)$ then $E|Z_0(t_1) - Z_0(t_2)|^2$ is proportional to $|t_1 - t_2|^{1+2\alpha}$.

Proof. As seen in Lemma 5 the part of the integrals away from the parameter values by an arbitrary $\epsilon > 0$ do not contribute to the limits of the Hellinger distances properly normalized. This implies that the contribution to the logarithm

$$\Lambda_n[\theta + \delta_n t; \theta] = \sum_{j=1}^{n} \{|X_j - \theta|^{\alpha} - |X_j - \theta - \delta_n t|^{\alpha}\}$$

that arise from values X_j such that $|X_j - \theta| > \epsilon$ will also tend to zero in probability. However, for $|X_j - \theta| < \epsilon$ and $|\delta_n t| < \epsilon$ one will have $||X_j - \theta|^{\alpha} - |X_j - \theta - \delta_n t|^{\alpha}| \leq 3\epsilon^{\alpha}$. This means, since ϵ can be taken arbitrarily small, that $\Lambda_n[\theta + \delta_n t; \theta]$ will always be asymptotically normally distributed. The same will be true of m-tuples $\Lambda_n[\theta + \delta_n t_i; \theta]$; $i = 1, 2, \ldots, m$ as explained in Chapter 5.

In addition, since the part of the integrals at distance $\geq \epsilon$ from θ do not contribute to the limits of normalized Hellinger distances, the Lindeberg conditions described in Chapter 5 are duly satisfied. In particular, the contiguity conditions will also be satisfied.

The form of the covariances for the limiting process are readily obtainable from the limits of normalized Hellinger distances described in Lemma 5. Hence the results. □

Note that here we have just claimed convergence of finite-dimensional distributions for the processes. Prakasa Rao [1968] has proved that the convergence takes place in the sense of Prohorov. Note also that in the case $\alpha = 1/2$ the limit has the familiar form that was described by the LAN conditions. They apply here as well as in the DQM_o case.

On the contrary, if $\alpha \in (0, 1/2)$, the LAN conditions are not satisfied and the limiting process is more complex. Its variables $Z(t)$ or $Z_0(t) = Z(t) - E[Z(t)]$ have for linear span an infinite-dimensional Hilbert space. Because of the relations between expectations and variances, the curve $t \rightsquigarrow EZ(t)$ is a helix whose linear span is also infinite-dimensional.

This is not surprising. It is a theorem that, *for the i.i.d. case, the only rates of convergence that can yield a limit process of the form $Z_0(t) = tW$ are the $\sqrt{n}$ rate or rates of the form $\varphi(n)\sqrt{n}$ where φ is slowly varying, as is $\sqrt{\log n}$.* See, for instance, Le Cam [1986, chapter 17, section 4].

It is a theorem all right, but not a good one. It uses the fact that if a sequence $\{a_n\}$ converges to a positive limit $a > 0$, then for each integer m the ratio a_{mn}/a_n must tend to one. This means that the result depends on the organization of the sequence as a sequence. The proof would not work if the sequence $\{n\}$ of integers was replaced by the sequence of prime numbers.

There are other results of this nature in the literature. The proof that a normed sum $S_n = a_n^{-1} \sum_{j=1}^{n} X_j - b_n$ with i.i.d. X_j's can have only stable distributions for limits depends on a similar argument. The usual argument about the possible limits of $a_n^{-1} \max_j [X_j; j \leq n] - b_n$ also falls in that category.

All of that falls apart if one lets the X_j; $j = 1, 2, \ldots, n$ be independent identically distributed with a distribution that depends on n.

For the case $\alpha \in (0, 1/2]$ the rates of convergence are faster than the usual $\sqrt{n}$ rate. One can wonder whether estimates exist that converge at these faster rates and could be used to construct centers as explained in Chapter 6. Estimates that converge at the appropriate rates do exist, as we shall see later (Section 7.5). They can be used to construct centers for $\alpha = 1/2$. For $\alpha \in (0, 1/2)$ one could attempt a construction of sorts, but the resulting "centers" are essentially unusable.

Example 3. This was communicated to us by J. B. H. Kemperman. It is given here to show that, even though cusps or peaks can change the asymptotic behavior, this need not be the case for multidimensional parameters.

Consider a density f with respect to the Lebesgue measure in $\Re^3$. Let f be of the form $f(x) = c|x|^{-1/2}\exp\{-|x|^2\}$ and take the corresponding shift family $\{p_\theta : \theta \in \Re^3\}$ where p_θ has density $f(x - \theta)$.

It is easily shown that this family satisfies DQM_o at all θ. Thus the asymptotics follow the LAN conditions of Chapter 6, with $\delta_n = n^{-1/2}$.

One can get worse examples. For instance, let $\{a_j; j = 1, 2, \ldots\}$ be an enumeration of the points with rational coordinates in $\Re^3$. Consider densities of the type

$$g(x - \theta) = c\{\sum_{j=1}^{\infty} 2^{-j}\sqrt{f(x - \theta - a_j)}\}^2$$

with f as above. Such a g has a dense forest of infinite peaks as θ varies. This does not affect the LAN conditions. The rate of convergence is still $\delta_n = n^{-1/2}$.

Note well, however, that in such a case the discretization of the auxiliary estimates used in Chapter 6 becomes very important. The auxiliary estimates θ_n^* and the points of the type $\theta_n^* + u_{n,i} + u_{n,j}$ must avoid the peaks.

Note also that here maximum likelihood estimates are rather poorly defined and poorly behaved.

Example 4. In the preceding example, we had infinite peaks, but we were in $\Re^3$. Similar examples can be constructed in $\Re^2$. In $\Re$ itself, however, an infinite peak will usually change the behavior of the family.

For instance, take a density f with respect to the Lebesgue measure on $\Re$. Let f be given by $f(x) = c|x|^{\alpha-1}e^{-|x|}$ where $0 < \alpha < 1$ is fixed. Let $\{p_\theta : \theta \in \Re\}$ be the corresponding shift family.

A computation analogous to that carried out for Example 2 shows that here, an appropriate rate of convergence is $\delta_n = n^{-\frac{1}{\alpha}}$. Here, as in Example 2 for small values of α, one can simplify one's life by using what we shall call Goria's principle: *If the rate of convergence δ_n is such that $\delta_n\sqrt{n} \to 0$ one can neglect a smooth part of the densities* (see Goria [1972]).

The asymptotic behavior here will not lead to gaussian processes. The log likelihood ratios will include terms proportional to $\sum_j \log(|X_j + \delta_n t|/|X_j|)$. The cases where the $|X_j|$ are of the same order of magnitude as $\delta_n t$ give a nonnegligible contribution. The condition $\max_j \log|(X_j + \delta_n t)/X_j| \to 0$ in probability is not satisfied.

For further developments, see Goria's [1972] thesis.

7.4 Some Nonparametric Considerations

In the preceding sections and Chapter 6, we have used families of probability measures indexed by subsets Θ of $\Re^k$. This indexing by itself does not mean much, but we also assumed that the structure of $\Re^k$ was in some way reflected by the measures. Using accepted terminology, this means that we looked at "parametric" situations. The reader should note that the terms "parametric" and "nonparametric" are not well defined. We shall suggest a possible definition in Section 7.5. Similar remarks can be made about the appellation "semiparametric," for which we shall not even attempt a definition.

In the present section we shall discuss ways of describing the local structure of an experiment that is not indexed by a subset of a euclidean space in such a way as to reflect the vector structure. Our first task will be to look at the paths and tangent cones of Pfanzagl and Wefelmeyer [1982]. Then we attempt to give a description of those experiments $\mathcal{E}^n$ formed from n independent identically distributed observations, with distributions possibly depending on n, singling out those that, as n tends to infinity, admit gaussian approximations. We limit ourselves to the case where pairs of measures do not separate entirely; see Conditions (B) and (N) of Section 5.3. For other cases, we only give references.

First, let us consider a set $\mathcal{P}$ of probability measures on a σ-field A. Recall that it defines an L-space, $L(\mathcal{P})$, of finite signed measures dominated by convergent sums $\sum a(p)p$ with $p \in \mathcal{P}$ and $\sum |a(p)| < \infty$; see Chapter 2.

It also defines a Hilbert space $\mathcal{H}(\mathcal{P})$ of symbols $\xi\sqrt{d\mu}$ where $\mu \geq 0$ is a measure in $L(\mathcal{P})$ and ξ is a function such that $\int \xi^2 d\mu < \infty$.

Two such symbols are given a squared (pseudo-)distance by

$$\|\xi_1\sqrt{d\mu_1} - \xi_2\sqrt{d\mu_2}\|^2 = \frac{1}{2}\int [\xi_1\sqrt{d\mu_1} - \xi_2\sqrt{d\mu_2}]^2.$$

They are considered identical if their distance is 0. It follows from this identification that in $\xi\sqrt{d\mu}$ one can always assume that μ is a probability measure, since one can always divide μ by its norm if it is not zero. Also, if μ and ν are mutually absolutely continuous, $\xi\sqrt{d\mu}$ is considered identical to $\eta\sqrt{d\nu}$ for $\eta = \xi\sqrt{d\mu/d\nu}$.

Addition is defined in the natural way. Given $\xi_i\sqrt{d\mu_i}$ one takes a measure m that dominates both μ_i and rewrites $\xi_i\sqrt{d\mu_i}$ as $\xi_i'\sqrt{dm}$. Then $\xi_1\sqrt{d\mu_1} + \xi_2\sqrt{d\mu_2} = (\xi_1' + \xi_2')\sqrt{dm}$.

To check the completeness of the space $\mathcal{H} = \mathcal{H}(\mathcal{P})$, take a Cauchy sequence $\{\xi_k\sqrt{d\mu_k}\}$, where the μ_k are probability measures. Introduce $\nu = \sum_k 2^{-k}\mu_k$, and rewrite the $\xi_k\sqrt{d\mu_k}$ as $\eta_k\sqrt{d\nu}$. The completeness follows from the completeness of the space of square integrable functions associated with ν.

If a $p \in \mathcal{P}$ is considered an element of $\mathcal{H}(\mathcal{P})$ it will usually be written $\sqrt{dp}$.

Call *path* in $\mathcal{P}$ any function $t \leadsto p_t$ from $[0, 1]$ to $\mathcal{P}$. If it is considered as function from $[0, 1]$ to the image of $\mathcal{P}$ in the Hilbert space, the path becomes $t \leadsto \sqrt{dp_t}$. The measure p_0 or the symbol $\sqrt{dp_0}$ will be called the endpoints of the path.

Definition 1. *A path $t \rightsquigarrow \sqrt{dp_t}$ will be called differentiable at zero if*

$$\frac{1}{t}[\sqrt{dp_t} - \sqrt{dp_0}]$$

tends to a limit in $\mathcal{H}$ as t tends to 0. That is, if there is a ξ and a μ such that

$$\int \{\frac{1}{t}[\sqrt{dp_t} - \sqrt{dp_0}] - \xi\sqrt{d\mu}\}^2$$

tends to zero.

It will be called a **Pfanzagl path** *if the limit has the form $v\sqrt{dp_0}$.* To put it differently, a Pfanzagl path is one that satisfies our condition DQM_o of Section 7.2 at $t = 0$.

Definition 2. *The Pfanzagl tangent cone T of $\mathcal{P}$ at p_0 is the set of derivatives at 0 of all Pfanzagl paths with endpoint p_0.*

That the said set of derivatives is a cone is easily seen, Indeed, for any $a > 0$ one can replace p_t by p_{at}. Our definition follows that of Pfanzagl and Wefelmeyer [1982, p. 22–25], but it looks very different. Pfanzagl's description of differentiability appears motivated by aspects needed for higher-order expansions. However, it happens to be equivalent to the one used for Pfanzagl paths here. The tangent cone was meant for the approximation of an experiment in the immediate vicinity of a given point. There are other entities that can sometimes be used for a similar purpose. Two of them are as follows. They do not use paths but sequences p_k or sequences of pairs (p_k, q_k).

Definition 3. *The contingent of $\mathcal{P}$ at p_0 is the cone of rays $\{\alpha v; \alpha \geq 0\}$ in which v is a limit (in $\mathcal{H}$) of ratios*

$$\frac{\sqrt{dp_k} - \sqrt{dp_0}}{\|\sqrt{dp_k} - \sqrt{dp_0}\|}$$

as $\|\sqrt{dp_k} - \sqrt{dp_0}\|$ tends to 0.

The paratingent of $\mathcal{P}$ at p_0 is the set of straight lines passing through 0 and a limit v of ratios

$$\frac{\sqrt{dp_k} - \sqrt{dq_k}}{\|\sqrt{dp_k} - \sqrt{dq_k}\|}$$

as both $\|\sqrt{dp_k} - \sqrt{dp_0}\|$ *and* $\|\sqrt{dq_k} - \sqrt{dp_0}\|$ *tend to* 0.

Contingents and paratingents were introduced by G. Bouligand in 1932. The contingents were used extensively in Saks [1932] *Theory of the Integral.* As defined, the contingent and the paratingent are cones with vertex 0, but the paratingent is symmetric around that vertex. It can be very much larger than the contingent, even if one symmetrizes the latter.

We shall also need *restricted* forms of these entities, restricted by admitting only for the limits involved in the definition of those that have the form $v\sqrt{dp_0}$, as was done for the Pfanzagl paths. It does not make any difference if one assumes, as is often done implicitly, that all the $p \in \mathcal{P}$ are absolutely continuous with respect to p_0.

Note that the tangent cones described here are quite different from the ones discussed in Section 7.2. Those were in the parameter spaces $\Re^k$; the present ones are in the Hilbert space $\mathcal{H}(\mathcal{P})$. They can readily be infinite-dimensional subspaces of the Hilbert space or the entire Hilbert space of symbols $\xi\sqrt{dp_0}$ where $\int \xi dp_0 = 0$ and $\int \xi^2 dp_0 < \infty$.

In Section 7.5, we shall discuss the use of these various tangent sets in establishing lower bounds for the risks of estimates. For now, let us indicate some of their relations with asymptotic normality.

Note that all three entities defined earlier are attached to an endpoint p_0. What follows assumes that endpoint is fixed. To get an asymptotic view, let us consider parametric families $\{p_t; t \in [0,1]\}$ and the corresponding families $\{P_{t,n}; t \in [0,1]\}$ of product measures, $P_{t,n}$ being the product of n copies of p_t.

In such a situation, the case of Pfanzagl paths was already discussed in Section 7.2, when we considered implications of the condition DQM_o. The asymptotic normality of the experiments $\{P_{s/\sqrt{n}}\}$ with s bounded is given by Proposition 1, Section 7.2.

One can do a bit better using a finite number of paths at the same time or using instead of $[0,1]$ a cube $[0,1]^k$ so that t becomes a vector in $\Re^k$. This would give a repeat of Section 7.2. Such a possibility is not easily available for the contingent cone, because the ratios occurring in the limits have denominators that are Hellinger distances $h(p_k, p_0) = \|\sqrt{dp_k} - \sqrt{dp_0}\|$, but one can still introduce product measures $P_{k,n}$ products of n copies of p_k. Consider then the case of the *restricted* contingent and of one sequence p_k. If it defines a ray of the restricted contingent, this ray may be written as positive multiples of a symbol $\xi\sqrt{dp_0}$, where ξ belongs to the space of p_0-square integrable functions. The relation that defines ξ may be written

$$\sqrt{dp_k} - \sqrt{dp_0} = h(p_k, p_0)[\xi\sqrt{dp_0} + \sqrt{dR_k}],$$

with the norm of the remainder term $\|\sqrt{dR_k}\|$ tending to zero. Multiplying this relation by $\sqrt{dp_0}$, integrating, and taking the limit, one sees that $\int \xi dp_0 = 0$. This implies, as in Proposition 1, Section 7.2 and Section 5.4, that if one takes n independent observations with n such that $nh^2(p_k, p_0)$ is bounded away from infinity, the experiments $(P_{k,n}, P_{0,n})$ are asymptotically gaussian. According to Section 5.4, this asymptotic gaussian character extends to experiments indexed by finite sets, implying the possibility of constructing a theory where one looks at several sequences, say, p_k^i as long as (1) they generate elements of the restricted contingent, and (2) the distances $h(p_k^i, p_0)$ tend to zero at the same general rate.

This would give a theory similar to what is available for the Pfanzagl tangent cones. By contrast the paratingents do not afford the possibility of such theories. Because they involve sequences of pairs (p_k, q_k) where $h(p_k, q_k)$ tends to zero and so does $h(p_k, p_0)$ but at possibly entirely different rates. To put it differently, take a number n of observations so that $nh^2(p_k, q_k)$ stays bounded and look at the product measures $(P_{k,n}, Q_{k,n})$. There is no reason that the behavior of these pairs is in any relevant way connected to that of $P_{0,n}$. The pairs in question should be studied on their own. A framework reflecting such intentions will be described shortly. But, we shall see in Section 7.5 that paratingents could be made to enter in extensions of the Barankin version of the Cramér-Rao bound.

The role played by the Pfanzagl tangent spaces cannot be appreciated unless one studies the so-called differentiable statistical functions. For these, see in particular Pfanzagl and Wefelmeyer [1982] and van der Vaart [1998].

Take any arbitrary set $\mathcal{P}$ of probability measures on a σ-field $\mathcal{A}$, without any restriction. Form the experiment $\mathcal{E}^n$ consisting of the measures P_n where P_n is the product of n copies of $p \in \mathcal{P}$ and the set of parameters is taken to be $\mathcal{P}$ itself or, better, the image of $\mathcal{P}$ in the Hilbert space $\mathcal{H}(\mathcal{P})$. One can inquire whether the experiments $\mathcal{E}^n$ are asymptotically gaussian. Here we have formulated the problem for a given $\mathcal{P}$, but it is more natural and informative to look at sets $\mathcal{P}_n$ that vary as n changes, giving rise to experiments $\mathcal{E}_n^n$. A particular case of such a variable situation occurred just above for binary experiments $(P_{k,n}, Q_{k,n})$. Similar situations were also discussed in Chapter 5. They arise most naturally in practice when one wants to study what appears to be an approximation to a complicated problem, simplifying it by neglecting some "not so important" aspects. What is considered "negligible" or "not so important" can and does usually depend on the information available.

Consider, for instance, a regression problem. If one has a total of 100 observations, one could set it in the form $Y_i = f(x_i) + \epsilon_i$ where there is only one explanatory variable with values x_i and where the ϵ_i are independent $\mathcal{N}(0, \sigma^2)$ variables. If, on the contrary, one has 10,000 observations, one might prefer a formulation $Y_i = g(u_i, x_i) + \epsilon_i$ where there are now two explanatory variables u_i and x_i and where with probability $(1 - \alpha)$ ϵ is $\mathcal{N}(0, \sigma^2)$ but with probability α, it suffers a contamination with large tail. Such changes of experiment from $n = 100$ to $n = 10,000$ are not uncommon. There is even some literature codifying the number of parameters to use. See, for instance, Akaike [1974] or G. Schwarz [1978].

In the regressions described, we have at least kept some semblance of common structure. One should not expect that much in general. Passages to the limit are often intended to avoid approximation arguments. Taking limits tends to emphasize certain aspects at the expense of the rest. They may produce simplifications, some of them unwarranted. Here, we have selected to underline the use of "triangular arrays" because their use is closer to approximation arguments and they do not prevent you from keeping some items fixed, but they do not suggest that this is what should be done. To be specific, consider an experiment $\mathcal{E}$ and its nth power $\mathcal{E}^n$. By nth power is meant that $\mathcal{E}$ is given by probability measures $\{p_\theta, \theta \in \Theta\}$. To get its nth power one repeats the experiment n times independently. Now suppose that for each integer k one has an experiment $\mathcal{E}_k$ and its powers $\mathcal{E}_k^n$ for $n = 1, 2, \ldots$. In such situations one can consider what happens in a particular row with k fixed. We shall call that the *power row* case. One can also consider what happens on the diagonal $k = n$ to the *diagonal array* of experiments $\mathcal{E}_n^n$. It has not been assumed that the parameter sets Θ_k are the same, independent of k, or that they have any connection to each other. Thus, as n varies, the parameter sets vary and so do the probability measures attached to them. We proposed to call that a "disarray" to emphasize the lack of assumed connections between the various experiments, $\mathcal{E}_n^n$, or, if one wants, as a contraction of "diagonal of the array."

The case of a power row $\mathcal{E}_k^n$ with $n = 1, 2, \ldots$ is the most common case studied in the literature. There, Θ_k is fixed and so are the measures p_θ. There are other cases, such as those that occur in local studies as in Section 2. In those, the p_θ are still fixed, but one makes the parameter spaces Θ_n vary, becoming smaller and smaller parts of Θ as n increases. Often, one "renormalizes" them.

Here our $\mathcal{E}_n^n$ is an nth-power of something. That is, it is obtained from n i.i.d. observations about which nothing much else is assumed. We have selected this formulation because it is closer to an attempt to study approximations and it does not prevent you from considering that certain items remain fixed, independent of n, if you so desire.

In the general diagonal array case, certain things may fail. For instance, the experiments $\mathcal{E}_n^n$ might not be approximable closely by their poissonized versions (see Le Cam [1986, p. 169]). There are also multitudes of examples where gaussian approximations do not work. (We have not proved that if gaussian approximations work, so does the poissonization, but it feels that way.)

Now let us come back to the general diagonal array case. A necessary condition for gaussian approximability was already given in Chapter 5. To state a form of it, consider a pair (p_n, q_n) of measures appearing in $\mathcal{E}_n^1$ whose nth-power yields $\mathcal{E}_n^n$. Form the likelihood ratio $r_n(p_n, q_n) = 2dp_n/(dp_n + dq_n)$ and, as before, let $h(p_n, q_n)$ be the Hellinger distance.

Lemma 7. *In order that there exist gaussian experiments $\mathcal{G}_n$ such that the distance $\Delta[\mathcal{E}_n^n, \mathcal{G}_n]$ tends to zero as n tends to infinity, it is necessary that for every pair (p_n, q_n) of elements of $\mathcal{E}_n^1$ such that $nh^2(p_n, q_n)$ stays bounded and for every $\tau > 0$ the quantity*

$$n(p_n + q_n)[|r_n - 1| \geq \tau]$$

tends to zero as n tends to infinity.

For this, see Section 5.4. As shown there, the condition implies gaussian approximability for sets $p_{i,n}$ of fixed finite cardinality as long as the boundedness of distances $\sqrt{n}h(p_{i,n}, p_{j,n})$ holds.

The necessary condition of Lemma 7 has a particular feature: It restricts only those pairs such that $nh^2(p_n, q_n)$ stays bounded. This is in the nature of pairs. Consider as in Lemma 7 a pair (p_n, q_n) of elements in the basic experiment $\mathcal{E}_n^1$. Their Hellinger affinity is $1 - h^2(p_n, q_n)$. The product of n copies of them will have an affinity

$$[1 - h^2(p_n, q_n)]^n \leq \exp -\{nh^2(p_n, q_n)\}.$$

If this tends to zero, the two product measures separate entirely. The experiment they yield can be approximated by a gaussian one, say, $\{\mathcal{N}(0,1), \mathcal{N}(2\sqrt{2n}h_n, 1)\}$, where we have written h_n for $h(p_n, q_n)$. The case of a finite set of pairs presents similar features. However, this does not cover the general situation, but it suggests looking more closely at the bounded case described by condition (B) as follows.

Condition B. *The array $\mathcal{E}_n^n$ satisfies condition (B) or is bounded if*

$$\sup_n \sup\{nh^2(p,q); p, q \in \mathcal{E}_n^1\} < \infty.$$

The condition is called (B) because of the label of a similar condition in Chapter 5. (Here, (B) implies the (N) of Chapter 5.)

Whether (B) holds or not, one can introduce for each n a Hilbert space H_n whose elements are the symbols $\xi\sqrt{dp}$ as before, with p occurring in the initial experiment $\mathcal{E}_n^1$ yielding the nth-power $\mathcal{E}_n^n$. However, we shall give them the square norm $\|\xi\sqrt{dp}\|_n^2 = 4n\int \xi^2 dp$. (This is intended to reflect the square distance of expectations in an approximating gaussian experiment.) It will be convenient to consider that the experiment $\mathcal{E}_n^n$ has for a set of parameters a certain subset A_n of H_n. Condition (B) is then the assumption that the diameter of A_n remains bounded independently of n.

As in Chapter 4, one can associate to H_n a canonical gaussian process, say, X, defined as a process $u \rightsquigarrow \langle u, X\rangle$ for u in H_n so that $E\langle u, X\rangle = 0$ and $E[\langle u, X\rangle]^2 = \|u\|^2$. This in turn defines a canonical gaussian experiment $\mathcal{G}$ indexed by H_n. We shall call $\mathcal{G}_n$ the experiment $\mathcal{G}$ but with parameter set restricted to A_n.

With such a notation, a possible question is: For what sets A_n does the distance $\Delta(\mathcal{E}_n^n, \mathcal{G}_n)$ tend to zero? Here the distance Δ is the distance between experiments defined in Chapter 2.

Definition 4. *The sequence $A_n \subset H_n$ will be called a gaussian aura if*

$$\Delta[\mathcal{E}_n^n(A_n), \mathcal{G}(A_n)]$$

tends to zero as n tends to infinity.

It was conjectured in Le Cam [1985] that *if the A_n satisfy condition (B)* the *necessary* condition of Lemma 7 is also *sufficient* for them to constitute a gaussian aura. Actually, the 1985 conjecture refers to a slightly more restricted situation to be described shortly, but it should extend to the present case. The conjecture was motivated by the following consideration. The distance Δ is defined by trying to match risk functions of the respective experiments as closely as possible and noting their maximum difference. One then takes the supremum of these differences over the entire class of decision problems, subject only to the restriction that the loss functions W used satisfy $0 \leq W \leq 1$.

One can get weaker distances by limiting the class of decision problems under consideration. One class with some appeal is a class of estimation problems as follows. Let $\mathcal{E}$ and $\mathcal{F}$ be two experiments with the same parameter set Θ. Let $\theta \leadsto \phi(\theta)$ be a function from Θ to the unit ball U of some Hilbert space. Consider the problem of estimating $\phi(\theta)$ with loss function $\|t - \phi(\theta)\|^2$ for the square of the norm of the Hilbert space. If an estimation procedure T_1 available on $\mathcal{E}$ yields a risk function $\theta \leadsto r(\theta, T_1)$, match it to a procedure T_2 available on $\mathcal{F}$ in such a way as to minimize $\sup_\theta \|r(T_1, \theta) - r(T_2, \theta)\|$. Proceed symmetrically, exchanging the roles of $\mathcal{E}$ and $\mathcal{F}$, then take the supremum of all such differences for all functions ϕ from Θ to U and all U. This will give a distance Δ_H.

It can be shown that, under condition (B), *the condition stated in Lemma 7 is necessary and sufficient for* $\Delta_H[\mathcal{E}_n^n, \mathcal{G}_n]$ *to tend to zero.*

Similar statements can be made for some other distances, such as Torgersen's distances Δ_k for k-decision problems and some distances obtainable from estimation problems in finite-dimensional spaces. For such statements, see Le Cam [1985].

One can use a different notation that may make the problem more intuitive and closer to the commonly described ones. It is as follows.

Take a set of probability measures, say, $\mathcal{P}_n$, and a particular element μ_n of it. Assume that all the elements of $\mathcal{P}_n$ are dominated by μ_n. Then an element p_n of $\mathcal{P}_n$ has a density $f_n = dp_n/d\mu_n$ whose positive square root yields $\int \sqrt{f}_n d\mu_n = 1 - h^2(p_n, \mu_n)$. Introducing the element $u_n = 2\sqrt{n}[\sqrt{f}_n - 1 + h^2(p_n, \mu_n)]$ of the space $\mathcal{L}_{2,0}(\mu_n)$ of functions u such that $\int u d\mu_n = 0$ and $\int \|u\|^2 d\mu_n < \infty$, this can be written in the form

$$\sqrt{f}_n = 1 - c(\frac{\|u_n\|}{2\sqrt{n}}) + \frac{u_n}{2\sqrt{n}} \tag{7.1}$$

where c is a certain function. For simplicity, we shall not make a notational distinction between a function and its equivalence class. For instance, we shall write that u belongs to the Hilbert space $L_{2,0} = L_{2,0}(\mu_n)$.

Expressing that f_n is a probability density, one gets that the function we have called c satisfies the identity

$$\int f_n \, d\mu_n = 1 = [1 - c(x)]^2 + x^2, \tag{7.2}$$

where we have written x for $\|u_n\|/2\sqrt{n}$. This identity defines c as a function from $[0, 1]$ to $[0, 1]$. Another identity can be obtained by writing $\sqrt{dp_n} = \sqrt{f_n}\sqrt{d\mu_n}$, multiplying by $\sqrt{d\mu_n}$, and integrating. It yields

$$c(\frac{\|u_n\|}{2\sqrt{n}}) = h^2(p_n, \mu_n), \tag{7.3}$$

as was clear from the preceding formulas. In the present case, f_n will be considered the density of an individual observation, that is, the density of a probability measure $p_n \in \mathcal{E}_n^1$, factor of the direct product, or power, $\mathcal{E}_n^n$. If that experiment was parametrized by the subset $A_n \subset H_n$, it can be reparametrized by the corresponding subset $A'_n \subset L_{2,0}$ of the various u_n.

Not every $u \in L_{2,0}$ can occur in the positive square root of a density. Since $\int f_n d\mu_n = 1$, one must have $\|u\|^2 \leq 4n$. Also, expressing that $\sqrt{f}_n \geq 0$ and introducing the negative part $u^- = \max[0, -u]$, one must have

$$|u^-|^2 \leq 4n[1 - c(\frac{\|u\|}{2\sqrt{n}}]^2 = 4n - \|u\|^2.$$

There is one simplification that can always be made without changing anything. *One can assume that $\mu_n = \mu$ independent of n and that it is the Lebesgue measure on a cube* $[0, 1]^\kappa$. Indeed, a dominated experiment is well defined by the joint distribution of its likelihood ratios to the dominating measure. It is a familiar fact, perhaps first observed by Paul Lévy, perhaps by P. J. Daniell, that for any sequence $X_r; r = 1, 2, \ldots$ of random variables there is another one, say, $X'_r; r = 1, 2, \ldots$ defined on $[0, 1]$ and having, for the laws induced by the Lebesgue measure the same distribution as the sequence X_r. The result extends to arbitrary families of random variables by taking a product $[0, 1]^\kappa$ with κ a suitable cardinal; see D. Maharam [1942]. Thus we can and will assume $\mu_n = \mu$ fixed.

With this simplification, the space $L_{2,0}(\mu)$ remains fixed. The subset $\mathcal{U}_n$ of those $u \in L_{2,0}$ that can occur in the positive square root of a density at the stage n increases as n increases. It will be called the set of *eligible* functions (at n). Any $u \in L_{2,0}$ that is bounded from below will be eligible for n large enough. However, the set of eligible functions has big holes. For instance, if μ is the Lebesgue measure, there are functions v_n such that $\|v_n\| = 1$ but $2\|v_n - u\| \geq 1 - (4\sqrt{n}/2^{2n})$ for any u eligible at n. By contrast, for a fixed $v \in L_{2,0}$ there are eligible u_n such that $\|v - u_n\|$ tends to zero as n tends to infinity.

Fixing the dominating measure μ puts us in a situation that resembles the situation of Section 7.2 or that used for describing Pfanzagl's tangent spaces, but it is different. The functions u_n are allowed to roam almost freely, so that, for instance, asymptotic normality is far from automatic and the experiments can change radically as n changes. To limit the possibilities we shall impose on them the strong condition (B). Here it can be rephrased by saying that there is a b independent of n such that $\|u_n\| \leq b$. A condition similar to the necessary condition for gaussian approximability, but one-sided, is as follows.

Condition L. *The experiments $\mathcal{E}_n^n$ with $u_n \in A'_n$ satisfy the uniform Lindeberg condition if for every $\tau > 0$*

$$\sup_{u \in A'_n} \int u^2 I[|u| \geq \tau\sqrt{n}]d\mu \tag{7.4}$$

tends to zero as n tends to infinity.

As previously mentioned, this condition is clearly *necessary* under (B) for gaussian approximability and it has been conjectured that it is also *sufficient.* We shall now recall some of the features it implies.

One of the implications is that one can replace the u_n by bounded functions. This is not a major result, but it is a technical one that we will use to obtain approximations that use exponential formulas. Specifically, for $u \in L_{2,0}$ and $\epsilon > 0$, let

$$u' = uI[|u| \leq \epsilon\sqrt{n}] - \int uI[|u| \leq \epsilon\sqrt{n}].$$

One can prove the following lemma in which $P_{u,n}$ is the distribution of n observations, each with density $[1 - c(\|u\|/2\sqrt{n}) + (u/2\sqrt{n})]^2$.

Lemma 8. *Assume $\|u\|^2 \leq b \leq 3n$ and $\int u^2 I[|u| > \epsilon\sqrt{n}]d\mu < \epsilon < 1/2$. Then if u is eligible, so is u' and $\|P_{u,n} - P_{u',n}\| \leq 2\sqrt{\epsilon}$.*

Now, introduce the empirical process S_n defined by

$$S_n(A) = \frac{1}{\sqrt{n}} \sum_{1 \leq j \leq n} [I_A(\omega_j) - \mu(A)]$$

where the ω_j are independent observations from the measure μ.

Denote $\langle u, S_n \rangle$ the integral $\int u dS_n$. One can define some other measures $Q_{u,n}$ as follows. Take for $Q_{0,n}$ the product of n copies of the basic measure μ and let

$$dQ_{u,n} = \exp\{\langle u, S_n \rangle - \frac{1}{2}\|u\|^2\} dQ_{0,n}.$$

This defines a family of measures. They are not necessarily all *finite* measures, unless the functions u are restricted as, for example, the u' of Lemma 8, but one has the following lemma.

Lemma 9. *Let $u \in L_{2,0}$ be eligible and such that $\|u\|^2 \leq b$. Assume, in addition, that $|u| \leq \epsilon\sqrt{n}$ with ϵ and $b\epsilon$ at most* $\log 2$.

Then $\|P_{u,n} - Q_{u,n}\| \leq 5\sqrt{b\epsilon}$.

The proof of the preceding two lemmas is a bit long, involving strings of inequalities, but no difficult or interesting; it shall not be given here (see Le Cam [1985]). However, there is a point that deserves mention. It is the point about uniform integrability of the densities in the definition of $Q_{u,n}$. It arises from a simple inequality on moment-generating functions. If X_j are independent with expectations zero. If $X_j \leq a$ and $\sum EX_j^2 = \sigma^2$ then, for $t \geq 0$, one has

$$E \exp[t \sum X_j] \leq \exp[\frac{1}{2} t^2 \sigma^2 e^{ta}].$$

There are also some inequalities involving the sum $\sum EX_j^2 I[|X_j| > \epsilon]$. We shall not insist, but we point out that, because of such inequalities, statements about small differences in distribution for, say, S_n and Z_n, translate immediately into small differences for expectations $E\{\exp[\langle u, S_n \rangle] - \exp[\langle u, Z_n \rangle]\}$.

The exponential expression occurring in the definition of our measures $Q_{u,n}$ is a familiar one. We encountered it in Chapter 4 in connection with gaussian experiments. Here the space $L_{2,0}$ introduces such an experiment naturally. Indeed, there exists a linear gaussian process, say, Z, attached to $L_{2,0}$ in such a way that $E\langle u, Z \rangle = 0$ and $E[\langle u, Z \rangle]^2 = \|u\|^2$. Denote G_0 the distribution of this centered process. Let G_u be that distribution shifted by u, so that G_u is the distribution of $u + Z$. Then

$$\frac{dG_u}{dG_0} = \exp\{\langle u, Z \rangle - \frac{1}{2}\|u\|^2\}.$$

It will be convenient to consider that the experiment $\mathcal{E}_n^n$ has for set of parameters a certain subset A_n of H_n. Condition (B) then is the assumption that the diameter of A_n remains bounded independently of n.

Now we have two families of measures, the $Q_{u,n}$ and the gaussian G_u. It is true that the $Q_{u,n}$ are not *probability* measures, but only close to them under conditions (B) and (L). However, they have log likelihood ratios given by *exactly* the same linear-quadratic expression $\langle u, x \rangle - (1/2)\|u\|^2$; at least it looks that way at first sight. A second look shows that this is illusory because S_n and Z take values in disjoint subsets of their space. The realizations of the variables $\langle u, S_n \rangle$ are discrete while the $\langle u, Z \rangle$ are continuous. In spite of this, the distribution for a given u of $\langle u, S_n \rangle$ is close to that of $\langle u, Z \rangle$ in the sense of the Prohorov distance or in the sense of the dual-Lipschitz distance. This is true for any finite subsets of $\mathcal{A}'_n$ as long as the cardinality of the finite subsets stays bounded.

These considerations suggest that one may be able to couple the processes S_n and Z on the same probability space in such a way that

$$\sup_u E|\exp\langle u, S_n\rangle - \exp\langle u, Z\rangle| \exp\{-\frac{1}{2}\|u\|^2\} \leq \eta, \tag{7.5}$$

where u ranges through $\mathcal{A}'_n$ and η is small, tending to zero as ϵ tends to zero.

Such a coupling procedure was suggested in Chapter 2 and would imply that the distance between our experiment $\mathcal{E}^n_n$, with $u \in \mathcal{A}'_n$, is close to a gaussian one (under conditions (B) and (L)). We have not succeeded in carrying out the coupling procedure in general. However, here is a theorem that uses a special case of it. It is stated for a reason that will be explained after the statement.

Theorem 1. *For each n, let $\{g_{n,k}, k = 1, 2, \ldots\}$ be a sequence of elements of $L_{2,0}$. Assume that $\|g_{n,k}\| \leq b$ for a fixed b and that the $g_{n,k}$ satisfy the uniform Lindeberg condition. Form the sums $\sum_k \alpha_k g_{n,k}$ with $\sum_k {\alpha_k}^2 \leq 1$. Let $\mathcal{A}_n$ be the set of such combinations that are eligible at n and let $\mathcal{E}^n_n$ be the resulting experiment.*

If the $g_{n,k}, k = 1, 2, \ldots$ are independent for each n, then the distance $\Delta(\mathcal{E}^n_n, \mathcal{G}_n)$ tends to zero as n tends to infinity.

We shall not reproduce the proof here, see Le Cam [1985]. It is a simple coupling made easy here by the independence of the $g_{n,k}$. One can do it for each value of n. One can even easily get crude bounds. For better bounds the technique of Komlós, Major and Tusnády [1975] can be used.

The reason we have stated Theorem 1 is that a coupling can often be obtained by using standard results on the convergence of empirical processes to gaussian ones in the sense of the Prohorov distance. For these, see the recent volume by R. M. Dudley [1999]. This kind of convergence requires some form of restriction on the Kolmogorov metric entropy of the sets $\mathcal{A}_n$. Here, no such restriction is imposed. The sets $\mathcal{A}_n$ are almost the entire unit ball of a Hilbert space. They do not satisfy the entropy restrictions needed for Prohorov-type convergence. One can combine the two approaches, but the statements so obtained are unwieldy and unappetizing. Let us pass instead to a different, more disturbing feature.

The results recalled above are relative to subsets of $L_{2,0}$ that stay bounded in norm. The individual distributions generating them are then at Hellinger distances $h(p,q) \leq b/\sqrt{n}$ for some b. Such neighborhoods are usable in a satisfactory manner in *parametric* studies. They are much too tiny for a proper study of *nonparametric* problems. In the latter, as a rule, it is not possible to estimate the probability measures with such precision and the interesting phenomena occur in much larger neighborhoods, for instance, of the type $h(p,q) \leq bn^{-\alpha}$ for α smaller or much smaller than $1/2$. (For a description of one of the controlling features, see Section 7.5.) Here is a result of M. Nussbaum that is not subject to this defect. It does not quite fit in the framework adopted so far because it is about experiments $\mathcal{E}^n$, not $\mathcal{E}^n_n$. However, it covers many important cases.

Theorem 2. *Consider an experiment $\mathcal{E}$ given by measures on $[0,1]$ that possess densities f with respect to Lebesgue measure. Assume that these densities satisfy a Hölder condition $|f(x) - f(y)| \leq |x-y|^\alpha$ for $\alpha > 1/2$ and a positivity condition $f(x) \geq \epsilon$ for some $\epsilon > 0$.*

Then, as n tends to infinity, $\mathcal{E}^n$ is asymptotically equivalent to the gaussian experiment generated by

$$dy(t) = \sqrt{f(t)}dt + \frac{1}{2\sqrt{n}}dW(t)$$

where W is standard Brownian motion.

The proof is rather complex and will not be reproduced here. We shall just mention some of its interesting features.

One of these consists in deriving the gaussian approximation on small neighborhoods $V(f,n)$ that shrink as n increases. The neighborhoods chosen are such that there exist estimates $\hat{f}_n$

with the property that, if f_0 is the true density, the probability under f_0, $P_0[\hat{f}_n \in V(f_0, n)]$ tends to unity. That forces them to be much larger than the balls $nh^2(p, p_0) \leq b$ considered earlier. It makes the proof of gaussian approximability on a $V(f, n)$ more complex. One of the techniques used by Nussbaum is to poissonize and split the poissonized experiment into components where likelihood ratios are tractable. On those, he uses a variant of the Komlos-Major-Tusnády approximation and a coupling approach. Then one has to put the poissonized pieces together.

To finish the proof, one pieces together the local gaussian approximations, (on the neighborhoods $V(f, n)$), using estimates based on only part of the sample.

There are some other results that have been proved for gaussian experiments. They might be extendable, with precautions, to our gaussian auras. One of them is a surprising inequality of Donoho and Liu [1991]. Suppose that the parameter set is suitably convex and take the loss function quadratic for simplicity. One then get a certain minimax risk, say, R, using all possible estimates. If instead one allows only *linear* estimates, one obtains a larger minimax risk, say, R_ℓ. The surprise is an inequality:

$$R_\ell \leq 1.247R.$$

To translate that for asymptotic purposes, one has to answer one question: *linear in what?* The question is more complicated than it seems. The variables of Donoho and Liu are the basic gaussian vectors in a Hilbert space. One could replace them by the log likelihood ratio process, suitably centered. In the case of our gaussian auras, this suggests using the empirical process, again suitably centered; see, Lemma 9. A similar argument on centered log likelihood processes can be applied to Nussbaum's neighborhoods or any situation where the coupling procedure of Section 2.4 can be used to derive gaussian approximations.

Another major result of Donoho refers to a "thresholding" procedure. A translation to the case where the experiments are only approximately gaussian is not simple and thus will not be given here.

Consider then the semiparametric problem of estimating the value $\langle \varphi, v \rangle$ of a linear functional φ defined and continuous on $\mathcal{L}_{2,0}(p)$ with the added information that v is already known to lie in a closed linear subspace L of $\mathcal{L}_{2,0}(p)$. Without this last information a standard estimate for $\langle \varphi, v \rangle$ would be $\langle \varphi, Z \rangle$. With the added information the standard estimate becomes $\langle \Pi\varphi, Z \rangle$ where Π is the orthogonal projection of φ on L.

Often the problem presents itself in a different way, involving "nuisance parameters." One is given a vector space W and a linear map from $\Re \times W$ to $\mathcal{L}_{2,0}(p)$, say, $(\xi, \eta) \rightsquigarrow v(\xi, \eta)$. One wants to estimate ξ while η is considered a nuisance. In such a case consider the level sets $L_\xi = \{v = v(\xi, \eta), \eta \in W\}$ where η is allowed to vary in W while ξ is kept constant. These level sets are linear affine subspaces of $\mathcal{L}_{2,0}(p)$, translates of the linear subspace $L_0 = \{v : v = (0, \eta), \eta \in W\}$. To get a simple problem with a simple answer we shall *assume that* L_0 *is a closed* subspace $\mathcal{L}_{2,0}(p)$ and that, *for* $\xi \neq 0$, L_ξ *is disjoint from* L_0. If so, the range $L = \{v : v = v(\xi, \eta),\ \xi \in \Re,\ \eta \in W\}$ is also closed in $\mathcal{L}_{2,0}(p)$ and there is some element φ of L that is orthogonal to L_0 and therefore to all the L_ξ. One can choose it so that $\langle \varphi, v \rangle = 1$ for all $v \in L_\xi$, $\xi = 1$. Then the problem of estimating ξ is reduced to the problem of estimating $\langle \varphi, v \rangle$, $v \in L$ as before.

Note that, in this situation, if we keep $\eta = 0$ and let ξ vary through $\Re$, the vector $v(\xi, 0)$ will trace a straight line in L. It can be written $v(\xi, 0) = \xi v(1, 0)$. Our φ is the projection of $v(1, 0)$ on the line of L orthogonal to L_0.

Stein [1956] has observed that the problem can be reduced to a two-dimensional one as follows. If $v(1, 0)$ is not orthogonal to L_0, let u be its projection on L_0. Then $\varphi = v(1, 0) - u$. By the independence of the $\langle v_j, Z \rangle$ for orthogonal v_j's, one can ignore all the $v \in L$ except those in the space spanned by u and $v(1, 0)$ or, equivalently, by u and φ.

This was described by Stein in a different manner. He took in W the one-dimensional subset that made the problem of estimation of ξ most difficult. He further suggested, as a matter of heuristics, that an estimate that solves this most difficult problem could be satisfactory for the initial problem.

In the gaussian situation described here, this is not a matter of heuristics. It is just so, but it becomes a matter of heuristics in the situations studied by Stein.

Still in the gaussian case, the estimate $\langle \varphi, Z \rangle$ of $\langle \varphi, v \rangle$ will have many optimality properties as one can judge from the results expounded in Chapter 5. The convolution theorem holds; the estimate has minimax properties for all bowl shape loss functions and so forth.

Note that the interference created by the nuisance parameter $\eta \in W$ has not been conjured away. If η was known, equal to zero, one would look at the experiments $\{G_v; v = \xi v(1,0), \xi \in \Re\}$ instead of $\{G_v; v \in L\}$. In these one-dimensional experiments, a shift from 0 to ξ induces a shift of the expectation of the normal variable whose square norm is $\xi^2 \|v(1,0)\|^2$. The corresponding shift in the projected experiments $\{G_v; v = \xi\varphi, \xi \in \Re\}$ is only $\xi^2\|\varphi\|^2$. The reduction coefficient is the cosine of the angle between $v(1,0)$ and φ.

The results given here depend explicitly on the fact that we are working on gaussian experiments indexed by a *linear* space and can use *linear* projections. They have been extended by Donoho [1990] and Donoho and Liu [1991]. In particular, these authors have extended C. Stein's result on the minimax property of the estimate for the most difficult one-dimensional problem to the case where the parameter space is only a *convex* subset in the basic Hilbert space.

Before we can apply analogous results to the asymptotic problems, we need to modify these results to a certain extent. There are some difficulties because our set A_n, set of functions $v \in \mathcal{L}_{2,0}(p)$ such that $\|v\|^2 \leq 4n$ and $(v^-)^2 \leq 4n - \|v\|^2$, does not contain any linear spaces, except the trivial space reduced to the origin. The gaussian theory based on full linear spaces such as L is not immediately applicable. Still, some statements are possible. Take a number b and consider the sets $\{v; v \in A_n \cap L, \|v\| \leq b\} = S(n,b)$.

There will be situations where these sets $S(n,b)$ will be a substantial part of the sets $L(b) = \{v; v \in L, \|v\| \leq b\}$. It may happen, for instance, that every $v \in L(b)$ is within an ϵ of a

$v' \in S(n, b)$.

In such a case, taking b large enough, one can still conclude that the minimax risks for parameters restricted to $S(n, b)$ will be very close to the minimax risk for L itself.

It is also necessary to contemplate other modifications. The functions called $v(\xi, \eta)$ here will become dependent on n, written $v_n(\xi, \eta)$, and they will not be exactly linear. However, there are many cases where they are quasi-linear in the sense that they are linear v_n^* such that $v_n(\xi, \eta) - v_n^*(\xi, \eta)$ will tend to zero uniformly for (ξ, η) bounded or, preferably, for $\|v_n\|$ bounded.

This means that one will have to deal with spaces L_n that are not linear but are close to linear and similarly with quasi-linear functions φ_n. It is a bit long to write in detail what happens, but it is clear that one can write down conditions that imply that the minimax properties are still preserved approximately, even if one restricts all considerations to balls $\{v; \|v\| \leq b\}$ provided that b is taken sufficiently large.

All of this can be done for the *gaussian* situation, but what does it say for our product experiments $\{Q_{v,n}; v \in B_n\}$ with $B_n = A_n \cap L_n$?

To see what happens, let us first select a large b and consider only the bounded sets $B_n(b) = (A_n \cap L_n) \cap \{\|v\| \leq b\}$.

For these one may check whether the uniform Lindeberg condition is satisfied. If so, for testing problems or for the one-dimensional estimation problems considered here, whatever is feasible for the gaussian $\{G_v; v \in B_n(b)\}$ is also feasible approximately for the true $\{Q_{v,n}; v \in B_n(b)\}$ and conversely.

If, on the contrary, the uniform Lindeberg condition is not satisfied, the theory expounded here does not volunteer any specific information on achievable risks.

In the case of estimation of linear or quasi-linear functions, further information can be obtained from the papers of Donoho and Liu [1991]. Some additional information will also be given in the next section.

The case of nonlinear functions is more complex; see the paper by Fan [1991] on the estimation of quadratic functionals.

7.5 Bounds on the Risk of Estimates

In this section we describe some lower bounds and some upper bounds on the risk of estimates. For lower bounds we have used a variant of the Cramér-Rao inequality according to some formulas related to the work of Kholevo [1973], Levit [1973], Pitman [1979], and others.

This is a formula that involves just two points in the parameter space. Formulas involving a finite number of possible parameter values but relying also only on characteristics of pairs of measures have been given by Fano (see Ibragimov and Has'minskii, [1981, p. 323]) and Assouad [1983]. They are also reproduced here but without proofs. (Proofs are given in Le Cam [1986].)

For upper bounds on minimax risks, we have given only one result, originally found in Le Cam [1975]. There are more recent results by Birgé [1983] that apply to independent observations that need not be identically distributed, but the proofs (see Le Cam [1986, p. 483–506] are somewhat more complex. Recently, many results have been obtained using the behavior of empirical processes on Vapnik-Červonenkis classes. In particular, Beran and Millar have used such techniques to construct confidence balls based on the half-space distance between measures on $\Re^k$.

One of the easy lower bounds on risks for quadratic loss makes use of an affinity number $\beta(P_0, P_1) = \int dP_0 dP_1/d(P_0 + P_1)$. If one takes densities f_0 and f_1 with respect to some dominating measure μ, this can also be written $\beta(P_0, P_1) = \int f_0 f_1/(f_0 + f_1)d\mu$. Consider then the problem of estimating a point $z \in [0, 1]$ with loss function $W_i(z) = (i - z)^2$, $i = 0, 1$. The estimate that minimizes the sum of the risks is easily seen to be $z = f_1/(f_0 + f_1)$. Its sum of risks is our number $\beta(P_0, P_1)$.

This number is related to a distance between measures $k(P_0, P_1)$ defined by

$$k^2(P_0, P_1) = \frac{1}{2} \int \frac{(dP_0 - dP_1)^2}{d(P_0 + P_1)} = 1 - 2\beta(P_0, P_1).$$

The same affinity number, or distance, can be used to write

an inequality that is more in line with the usual Cramér-Rao lower bound. For this purpose, consider a real-valued estimate T. One would like to have differences $E_0T - E_1T$ large while maintaining small variances.

Lemma 10. *Any estimate T always satisfies the inequality*

$$var_0 T + var_1 T \geq (E_0T - E_1T)^2 \frac{\beta(P_0, P_1)}{1 - 2\beta(P_0, P_1)}$$

where $var_i T$ is the variance of T for P_i.

Note 1. To compare this with the Cramér-Rao inequality, it is better to rewrite it in the form

$$var_0 T + var_1 T \geq [E_0T - E_1T]^2 \frac{1}{2} \frac{[1 - k^2(P_0, P_1)]}{k^2(P_0, P_1)}.$$

Proof. Let $m = (1/2)(E_0T + E_1T)$ and $\Delta = E_0T - E_1T$. Then

$$var_0 T + var_1 T = E_0(T - m)^2 + E_1(T - m)^2$$

$$-(E_0T - m)^2 - (E_1T - m)^2$$

$$= (E_0 + E_1)(T - m)^2 - \frac{1}{2}\Delta^2.$$

Taking densities f_i with respect to the measure $S = P_0 + P_1$, this may be written

$$\int (T - m)^2 dS - \frac{1}{2}[\int (T - m)(f_0 - f_1) dS]^2.$$

By Schwarz's inequality this will be as small as possible, for a given Δ, if $T - m$ is of the form $T - m = c(f_0 - f_1)$. Then $\Delta = c \int (f_0 - f_1)^2 dS$, thus fixing the value of c. Substituting back in the expression for the sum of variances gives the result. □

Unfortunately this inequality is difficult to use when the measures P_i are product measures. Thus one uses another inequality where our number β has been replaced by the Hellinger affinity $\rho(P_1, P_0) = \int \sqrt{dP_1 dP_0}$. This multiplies happily in product situations and gives the following.

Corollary 1. *Any estimate T satisfies the inequality*

$$var_0 T + var_1 T \geq [E_0 T - E_1 T]^2 (1/2) \frac{\rho^2(P_0, P_1)}{[1 - \rho^2(P_0, P_1)]}.$$

Indeed the right term in the inequality of Lemma 10 is an increasing function of β and $\beta(P_0, P_1) \geq \rho^2(P_0, P_1)/2$.

Note that the proof of Lemma 10 shows that, given a particular pair (P_0, P_1), there exists an estimate that achieves equality in the formula of Lemma 10. This is no longer the case for the inequality in the corollary unless the ratio $|f_1 - f_0|/(f_1 + f_0)$ takes only one value, since this is the only case in which $\beta = \rho^2/2$. Either Lemma 10 or its corollary can be used to obtain some form of the Cramér-Rao inequality by taking sequences $\{P_{1,n}\}$ where $h(P_0, P_{1,n})$ tends to zero.

The Cramér-Rao inequality is perhaps the most famous and most used lower bound for the risk of estimates. We shall give here a derivation that, although well known, has not appeared much in standard textbooks, because the proof that the bound it yields is attained uses some infinite-dimensional vector space theory. We follow Barankin [1949] who sets the problem in a neat framework.

Let $\mathcal{E} = \{P_\theta; \theta \in \Theta\}$ be an experiment, and let $\theta \rightsquigarrow \gamma(\theta)$ be a numerical function defined on Θ. Look for estimates T available on $\mathcal{E}$ such that $\int T dP_\theta = \gamma(\theta)$ identically for $\theta \in \Theta$. Among those, try to find some that are most "likeable" in a sense to be made precise. For instance, one may specify a particular θ_0, let $P = P_{\theta_0}$ and declare that the most "likeable" estimate is one for which $\int T^2 dP$ is the smallest possible. There are other criteria; some will be mentioned later.

Introduce the space $\mathcal{M}$ of finite signed measures with finite support on Θ and let $P_\mu = \int P_\theta \mu(d\theta)$. The map $\mu \rightsquigarrow P_\mu$ is a linear map onto a certain subspace $\mathcal{S}$ of finite signed measures on the σ-field that carries $\mathcal{E}$. The requirement that $\int T dP_\theta \equiv \gamma(\theta)$ implies $\int \gamma(\theta)\mu(d\theta) \equiv \int T dP_\mu$ for all $\mu \in \mathcal{M}$. To simplify writing we shall let $\langle \gamma, \mu \rangle = \int \gamma(\theta)\mu(d\theta)$ for the linear functional so defined.

The Cramér-Rao inequality then takes the following form.

Proposition 4. *Assume that the P_θ are dominated by P and the densities $f_\theta = dP_\theta/dP$ are square integrable for P.*

Then any estimate T such that $\int T dP_\theta \equiv \gamma(\theta)$ for all $\theta \in \Theta$ must also satisfy the inequality

$$\int T^2 dP \geq \sup_\mu \frac{|\langle \gamma, \mu \rangle|^2}{\int \frac{[dP_\mu]^2}{dP}}.$$

Proof. This follows from the Schwarz inequality. Since $\langle \gamma, \mu \rangle = \int T f_\mu dP$ with $f_\mu = dP_\mu/dP$, one has

$$|\langle \gamma, \mu \rangle|^2 \leq \int T^2 dP \int f_\mu^2 dP.$$

The inequality is often stated in a different way, in terms of variances. It then takes the following form.

Proposition 4a. *Under the conditions of Proposition 4 one has*

$$\mathrm{varT}_P \geq \sup_{\mu,\nu} R(\mu.\nu)$$

where the supremum is taken over μ and ν, probability measures in $\mathcal{M}$, and where

$$R(\mu, \nu) = \frac{|\langle \gamma, \mu - \nu \rangle|^2}{\int \frac{|dP_\mu - dP_\nu|^2}{dP}}. \tag{7.6}$$

This is obtained by algebraic manipulations on the expression of Proposition 4 together with the remark that if $\int T dP = \gamma(\theta_0) = 0$, one has $\langle \gamma, \mu \rangle = \langle \gamma, \mu + c\delta \rangle$ for c real and δ the Dirac mass at P.

Note that the ratio $R(\mu, \nu)$ is not changed if one simultaneously replaces μ by $\epsilon\mu + (1-\epsilon)\delta$ and ν by $\epsilon\nu + (1-\epsilon)\delta$. Thus, the inequality can be obtained by considering what happens in arbitrarily small neighborhoods of P in the convex hull of the P_θ's, where the pairs (μ, ν) lie, or in the linear space $\mathcal{S}$ they span.

Barankin's [1949] argument differs from the preceding. It has additional merits, among which one can cite the following:

1. Under the conditions of Proposition 4, that is, domination by P and square integrability of the densities, the argument yields *existence* of estimates T that achieve the bounds of Proposition 4 or 4a. In particular this shows that the bounds are "the best of their kind," as Fréchet would have said.

2. The argument is not restricted to square norms or variances. It applies, for example, to L_p-norms for $1 < p < \infty$ and to minimax values of risks.

3. The condition that $\int T dP\theta \equiv \gamma(\theta)$ exactly is a strong condition that may be imposed for mathematical purposes. For practical use, one may prefer conditions of the type

$$|\int T dP_\theta - \gamma(\theta)| \leq \epsilon(\theta),$$

where $0 < \epsilon(\theta) < \infty$. Barankin's argument is adaptable to this case and many variations on it.

We shall not present Barankin's argument in full, but we'll try to indicate some of its basic features. It relies on a small bit of linear space theory, the bipolar theorem, itself an easy consequence of the Hahn-Banach theorem. One can start as follows. Let $\mathcal{S}$ be the space of linear combinations $P_\mu = \int P_\theta \mu(d\theta)$ with $\mu \in \mathcal{M}$. Assume for simplicity that $P_\mu = 0$ implies $\mu = 0$. This by itself is not essential but will allow us to dispense with several maps and identify $\mathcal{S}$ and $\mathcal{M}$ for linear purposes. The identification has a (at first unintended, but important) consequence: The functional $\langle \gamma, \mu \rangle = \int \gamma(\theta)\mu(d\theta)$ originally defined on $\mathcal{M}$ is also a linear functional on $\mathcal{S}$. If $P_\mu = \int P_\theta \mu(d\theta) = 0$ did not imply $\langle \gamma, \mu \rangle = 0$ there could not be any statistic T such that $\int T dP_\theta \equiv \gamma(\theta)$.

Let us suppose that a certain set C of estimates has been selected as a standard of quality. Assume that $T \in C$ implies $\int |T| dP_\theta < \infty$ for all θ. Each $T \in C$ defines a linear functional $\langle \phi_T, \mu \rangle$ on $\mathcal{S}$ and $\mathcal{M}$ by

$$\langle \phi_T, \mu \rangle = \int T dP_\mu.$$

Let $\mathcal{S}'$ be a set of linear functionals defined on $\mathcal{S}$. Assume that this is a linear space and that it contains the functional defined by γ and all the functionals defined by C. Identify C with the set of functionals defined by C.

By definition, the polar C^0 of C in $\mathcal{M}$ (or $\mathcal{S}$) is the set of μ such that $\langle f, \mu \rangle \leq 1$ for all $f \in C$. Similarly, the bipolar of C in $\mathcal{S}'$ is the set C^{00} of $f \in \mathcal{S}'$ such that $\langle f, \mu \rangle \leq 1$ for all $\mu \in C^0$. The bipolar theorem says:

If C is convex, contains 0, and C is closed for the weak topology of pointwise convergence on $\mathcal{M}$, then $C = C^{00}$.

From this it follows readily that if $C = C^{00}$, there is an $r \in [0, \infty)$ such that $\gamma \in rC$ if and only if $\langle \gamma, \mu \rangle \leq r$ for all $\mu \in C^0$. In other words, there is an estimate $T \in rC$ such that $\int T dP_\theta \equiv \gamma(\theta)$ if and only if $\langle \gamma, \mu \rangle \leq r$ for all μ in the polar of C. There is then a smallest r for which that is true. It can be taken as an index of quality, or more exactly, lack of quality of T.

We have mentioned that the method is applicable to minimax variance. One just takes the appropriate C. If one does so, one sees that the minimax variance involves quantities such as

$$\frac{|\langle \gamma, \mu \rangle - \langle \gamma, \nu \rangle|^2}{\int \frac{(dP_\mu - dP_\nu)^2}{dP_\lambda}}$$

with $\lambda = (\mu + \nu)/2$ as already encountered in Lemma 10.

In fact, it is possible to obtain the value of the minimax variance through the following steps in whose description we let $\lambda = (\mu + \nu)/2$ for μ and ν that are probability measures elements of $\mathcal{M}$. We also write

$$K^2(\mu, \nu) = (1/2) \int \frac{(dP_\mu - dP_\nu)^2}{d(P_\mu + dP_\nu)},$$

as was done for the k^2 of Lemma 10.

(a) Fixing μ and ν, take the minimum of the variance under P_λ of an estimate T subject to the sole restriction that $\int T(dP_\mu - dP_\nu) = \langle \gamma, (\mu - \nu) \rangle$. This gives a value

$$V(P_\lambda) = \frac{(\langle \gamma, \mu \rangle - \langle \gamma, \nu \rangle)^2}{4K^2(\mu, \nu)}.$$

(b) Keeping $\lambda = (\mu + \nu)/2$ fixed, take the supremum over μ and ν. This gives a value, say, $R(\lambda)$.

(c) Take the supremum over λ of

$$R(\lambda) - v(\gamma; \lambda)$$

where $v(\gamma; \lambda)$ is the variance of γ under λ, that is, $v(\gamma; \lambda) = \int \gamma^2 d\lambda - (\int \gamma d\lambda)^2$.

Note that the only place where a minimum over T occurs is in the first step. It involves only two measures P_μ and P_ν and is very reminiscent of Lemma 10.

There are other inequalities that use finite sets Θ with more than two points. One of them uses the Kullback-Leibler information numbers $K(P, Q) = \int(\log(dP/dQ))dP$. With $J(P, Q) = K(P, Q) + K(Q, P)$, it reads as follows.

Lemma 11. (Fano). *Let Θ be a set of cardinality $n > 2$. Consider the problem of estimating θ with a loss function W such that $W(s, t) = 1$ if $s \neq t$ and $W(s, s) = 0$. Then the minimax risk is at least*

$$1 - \frac{1}{\log(n-1)}[\log 2 + \frac{1}{2} \max_{j,k} J(P_j, P_k)].$$

This lemma does not require any particular structure for the set Θ. Another lemma, due to Assouad, assumes that Θ is a "hypercube" of dimension r, set of sequences $\theta = \{\xi_1, \xi_2, \ldots, \xi_r\}$, where ξ_j takes value 0 or 1. For two such sequences, say, θ and θ', let $W(\theta, \theta')$ be equal to the number of coordinate places where the sequences differ.

Lemma 12. (Assouad). *Let Θ be a hypercube of dimension r. Let W be the loss function just defined. Then the minimax risk is at least*

$$\frac{r}{2} \inf\{\|P_s \wedge P_t\|; (s, t) \in \Theta \times \Theta, W(s, t) = 1\}.$$

We shall not prove these results here; for a proof see for instance Le Cam [1986]. Our reason for mentioning them is this:

Even though the results are for finite sets Θ with more than two points, they make use only of affinities or distances between pairs (P_s, P_t). This makes them relatively easy to use. However, the difficulty of an estimation problem is not entirely described by the behavior of pairs. That the results can still be very useful can be judged by the works of Ibragimov and Has'minskii [1981], Bretagnolle and Huber [1979], and Birgé [1983].

Another attack on estimation bounds has been carried out using results on tests. Suppose, for instance, that one wishes to estimate a real-valued function g on a set Θ and measures $\{P_\theta; \theta \in \Theta\}$. Suppose that there exist estimates T and numbers α and δ such that $P_\theta\{|T - g(\theta)| \geq \delta\} \leq \alpha$ for all $\theta \in \Theta$. Consider two possible values a and b of g. Assume $b \geq a + 2\delta$. Then T will provide a test of $A = \{P_\theta; g(\theta) \leq a\}$ against $B = \{P_\theta; g(\theta) \geq b\}$ by accepting the first set or the second according to whether $T \leq (a+b)/2$ or $T > (a+b)/2$. Such a test φ will have a sum of errors $\int \varphi dP_s + \int (1-\varphi) dP_t \leq 2\alpha$ for all $\{s : g(s) \leq a\}$ and $\{t : g(t) \geq b\}$. This suggests introducing a measure of affinity as follows.

Let A and B be two sets of measures on a given σ-field. For any test function φ, let

$$\pi(A, B; \varphi) = \sup_{s,t}\{\int \varphi dP_s + \int (1-\varphi) dP_t; s \in A, t \in B\}.$$

Let $\pi(A, B) = \inf_\varphi \pi(A, B; \varphi)$. This will be called the *testing affinity* between A and B.

The very definition of $\pi(A, B; \varphi)$ shows that it is not affected if one replaces A by its convex hull con A, and B by its convex hull, con B. It is also not affected if one performs a closure operation as follows. Let M be the space of bounded measurable functions. Consider the measures as elements of the dual M' of M. Close con A and con B in M' for the weak topology $W(M', M)$ induced by M. Let $\hat{A}$ and $\hat{B}$ be these closed convex sets.

It can be shown that $\pi(A, B)$ *is precisely*

$$\sup\{\|P \wedge Q\|; P \in \hat{A}, Q \in \hat{B}\}.$$

(Le Cam [1986] uses an extended definition of "tests." For that extended definition it is enough to take the convex closures in the initial space of σ-additive measures for the topology induced by the L_1-norm. The two definitions coincide if the families A and B are dominated by a σ-finite measure.) This is a simple consequence of the Hahn-Banach theorem. A proof is given in Le Cam [1986, p. 476].

In such statements the closure operation rarely poses serious problems. The difficulties occur in the convexification operation, going from A to con A and the same for B.

Consider, for instance, the case where the set A is $A_n = \{P_{\theta,n}; g(\theta) \leq a\}$ for product measures $P_{\theta,n}$, distributions of n independent variables with common distribution p_θ. It is hard to tell on the individual p_θ what will be in the convex hull of A_n. A typical procedure to evaluate distances between con A_n and con B_n with $B_n = \{P_{\theta,n}; g(\theta) \geq b\}$ will be to try to find a finite family $\{\varphi_j; j \in J\}$ of bounded measurable functions such that for each pair $s \in A_n$ and $t \in B_n$ there is at least one $j \in J$ with $\int \varphi_j dp_s < \int \varphi_j dp_t$. If that can be done, the two convex hulls con A_n and con B_n will separate for sufficiently large n; see Le Cam and Schwartz [1960]. When that happens, the separation then proceeds exponentially fast. That is, the testing affinity decreases exponentially fast. For applications of this to lower and upper bounds on the rates of convergence of estimates of linear functions of the p_θ, see Donoho and Liu [1991].

So far, we have considered only arguments that yield lower bounds for risks. There are also arguments that yield upper bounds. We shall just give one as an example. It is taken from Le Cam [1973, 1975]. For further results, see Birgé [1983].

Consider first a set $\{P_\theta; \theta \in \Theta\}$ of probability measures P_θ on a σ-field $\mathcal{A}$. (Later, the P_θ will acquire subscripts "n" as usual.) Let W be a metric on Θ. Any subset A of Θ has a diameter, diam $A = \sup\{W(s,t);\ s \in A,\ t \in A\}$. Take two sequences $\{b_\nu; \nu = 0, 1, 2 \ldots\}$ and $\{a_\nu; \nu = 0, 1, 2, \ldots\}$ such that $b_\nu > b_{\nu+1}$ and $b_\nu > a_\nu > 0$ for all ν. For each ν, let $\{A_{\nu,i}; i \in I_\nu\}$ be a partition of Θ by set $A_{\nu,i}$ whose diameter does not exceed a_ν.

Assumption 1. *The diameter of Θ does not exceed $b_0 < \infty$. The partition $\{A_{\nu,i}; i \in I_\nu\}$ has finite cardinality. It has the* *minimum possible number of sets among all partitions by sets whose diameter does not exceed* a_ν.

Using such a sequence of partitions, we shall construct a decreasing sequence $\{S_k; k = 1, 2 \ldots\}$ of random sets such that the diameter of S_k does not exceed b_k. They are intended to play the role of confidence sets.

The probabilities $P_\theta[\theta \notin S_k]$ will increase as k increases. Thus, to keep these probabilities small, we shall stop k at some integer m depending on the structure of Θ. Bounds on the probabilities are given in Proposition 5 in terms of the numbers a_ν and b_ν, the cardinalities of the partitions, and the error probabilities of tests.

Further bounds, in terms of a dimension function for Θ, are given in Proposition 6. One obtains estimates $\hat{\theta}$ by taking any point in the last (nonempty) retained S_k. Then, from bounds on $P_\theta[\theta \notin S_k]$, one can readily obtain bounds on the risk $E_\theta W^2(\hat{\theta}, \theta)$. It will be seen that these bounds and the dimension function are intimately related to the rates of convergence of estimates.

To proceed call a pair (i, j) of elements of I_ν, or the corresponding pair $(A_{\nu,i}, A_{\nu,j})$ in the partition, a b_ν-distant pair if

$$\sup_{s,t}\{W(s,t);\ s \in A_{\nu,i},\ t \in A_{\nu,j}\} \geq b_\nu.$$

For such a b_ν-distant pair, let nonrandomized $\varphi_{\nu,i,j}$ be a test of $A_{\nu,j}$ against $A_{\nu,i}$ selected so that $\varphi^2_{\nu,i,j} = \varphi_{\nu,i,j} = 1 - \varphi_{\nu,j,i}$.

Construct sets $S_0 \supset S_1 \supset S_2 \ldots$ as follows. Let $S_0 = \Theta$. If $S_0, S_1, \ldots, S_{\nu-1}$ have been constructed, let $J_\nu \subset I_\nu$ be the set of indices $i \in I_\nu$ such that $A_{\nu,i}$ intersects $S_{\nu-1}$. If $i \in J_\nu$, let $J_\nu(i)$ be the set of $j \in J_\nu$ that are b_ν-distant from i. Perform a test of $A_{\nu,j}$ against $A_{\nu,i}$ by $\varphi_{\nu,i,j}$ and eliminate the rejected set. Let $\psi_{\nu,i}$ be the test function $\psi_{\nu,i} = \inf_j\{\varphi_{\nu,i,j}; j \in J_\nu(i)\}$ so that $\psi_{\nu,i} = 1$ if and only if $A_{\nu,i}$ is accepted in all pairwise tests. For each sample point ω there is a certain set of indices $i \in J_\nu$ such that $\psi_{\nu,i}(\omega) = 1$. Let it be $R_\nu(\omega)$. Then let

$$S_\nu = \bigcup_i \{A_{\nu,i} \cap S_{\nu-1};\ i \in R_\nu(\omega)\}.$$

The process stops when $R_\nu(\omega)$ becomes empty, but it can also be stopped when $\nu = m$ for a prescribed integer m.

That procedure may look complicated, but all it amounts to is at each stage testing only between distant pairs that intersect $S_{\nu-1}$ and retaining only those members of the pairs $(A_{\nu,i}, A_{\nu,j})$ that are accepted by all the tests.

An indication of the performance of the procedure can be obtained from the testing affinity defined by

$$\pi_\nu(i,j) = \inf_{\varphi} \sup_{s,t} [\int (1-\varphi) dP_s + \int \varphi dP_t],$$

where $s \in A_{\nu,i}$, $t \in A_{\nu,j}$, and φ is a test function. For any pair (i, j) we can assume that one has taken $\varphi_{\nu,i,j} = \varphi^2_{\nu,i,j}$ so that

$$2\pi_\nu(i,j) \geq \sup_{s,t} [\int (1-\varphi_{\nu,i,j}) dP_s + \int \varphi_{\nu,i,j} dP_t]$$

for a supremum taken for $s \in A_{\nu,i}$ and $t \in A_{\nu,j}$. If such a test did not exist one would take an $\epsilon > 0$ and tests that achieve the inequality within an ϵ. This would complicate the argument without changing the final results to any material extent. Assuming existence of the tests, one gets the following result.

Proposition 5. *Let Assumption 1 hold. Assume also that there are numbers $\beta(\nu)$ such that $\pi_\nu(i,j) \leq \beta(\nu)$ for all distant pairs at ν. Let $U_\nu(\theta)$ be the ball of radius $a_\nu + b_{\nu-1}$ centered at θ. Assume that there is a number $C(\nu)$ such that every $U_\nu(\theta)$ can be covered by at most $C(\nu)$ sets of diameter a_ν or less. Then, for every $k \geq 1$ one has diam $S_k \leq b_k$ and*

$$P_\theta[\theta \notin S_k] \leq 2 \sum_{\nu=1}^{k} \beta(\nu) C(\nu).$$

Proof. The assertion diam $S_k \leq b_k$ is a consequence of the fact that, at each step, one of the sets $A_{\nu,i}$ or $A_{\nu,j}$ of a distant pair is eliminated. For the probabilistic assertion, fix a particular $\theta \in \Theta$.

Let J_ν^* be the set of indices $j \in I_\nu$ such that $A_{\nu,j}$ intersect $V_\nu(\theta) = \{t \in \Theta; W(\theta,t) < b_{\nu-1}\}$. For all $j \in J_\nu^*$, the $A_{\nu,j}$ are contained in the ball $U_\nu(\theta)$ of radius $a_\nu + b_{\nu-1}$. Thus there cannot be more than $C(\nu)$ of them. Indeed suppose that there would be $N_\nu \geq C(\nu)+1$ of them; then one could cover $U_\nu(\theta)$ by a certain number $N' < N_\nu$ of sets $A'_{\nu,i}$ such that diam $A'_{\nu,i} \leq a_\nu$. One can assume that they form a partition of $U_\nu(\theta)$. Those, together with the $A_{\nu,i}$ that do not intersect $V_\nu(\theta)$, will yield a partition of Θ by a number of sets strictly inferior to that of the partition $\{A_{\nu,i}; i \in I_\nu\}$. Since this one was supposed to be minimal, this is a contradiction.

Now there is one set of the partition $\{A_{\nu,i}; i \in I_\nu\}$ that contains θ. For simplicity of notation call it $A_{\nu,0}$. This set will be tested at most against all the $A_{\nu,j}$ that (1) intersect $V_\nu(\theta)$ and (2) are distant at stage ν from $A_{\nu,0}$. There cannot be more than $C(\nu)$ of them and their individual probability of rejecting $A_{\nu,0}$ is at most $2\pi_\nu(i,j)$. The result follows by addition of the probabilities. Hence the proposition. □

To construct an estimate $\hat{\theta}$ of θ one can choose any integer m and any point $\hat{\theta}$ in the last S_k, $k \leq m$ that is not empty. This gives the following corollary.

Corollary 2. *Let g be any monotone increasing function from $[0,\infty)$ to $[0,\infty)$. Define an estimate $\hat{\theta}$ as stated above then*

$$E_\theta g[W(\hat{\theta},\theta)] \leq g(b_m) + 2 \sum_{0 \leq \nu \leq m-1} g(b_\nu)\beta(\nu+1)C(\nu+1).$$

This follows from the above by an algebraic rearrangement akin to "integration by parts." using the fact that the $g(b_\nu)$ decrease as ν increases.

Proposition 5 can be applied to any metric W. It is also valid for pseudometrics. In fact, one could extend it further, but it will work reasonably only for functions W such that pairs $A_{\nu,i}, A_{\nu,j}$ are "distant" already implies the existence of good tests.

One could use it with distances such as the Kolmogorov–Smirnov distance, or the half-space distance, but for those it is simpler to use minimum distance estimates. We shall apply it instead to Hellinger distances.

To do this, consider measures $P_{\theta,n}$, $\theta \in \Theta$, distributions of n independent observations with common individual distribution p_θ. Let h be the Hellinger distance for one observation, so that

$$h^2(s,t) = \frac{1}{2}\int(\sqrt{dp_s} - \sqrt{dp_t})^2.$$

Let $W^2(s,t) = nh^2(s,t)$. Take a sequence $b_\nu = b_0 q^\nu$, where q is some number $q \in (0,1)$. Let $a_\nu = b_\nu / r$, where r is some number $r > 1$.

If two sets $A_{\nu,i}$ and $A_{\nu,j}$ are distant at stage ν, there will be points (s,t) with $s \in A_{\nu,i}$ (or its closure) and $t \in A_{\nu,j}$ (or its closure) such that $W(s,t) \geq b_\nu$ and $A_{\nu,i}$ is contained in the ball of center s and radius a_ν for W. Similarly $A_{\nu,j}$ will be in a ball of center t and radius a_ν. They will be disjoint if $b_\nu > 2a_\nu$, that is, if $r > 2$.

Let us look also at the sets $A^1_{\nu,i} = \{p_\theta; \theta \in A_{\nu,i}\}$ and $A^1_{\nu,j} = \{p_\theta; \theta \in A_{\nu,j}\}$ in the space of measures as metrized by the Hellinger distance h. Since $W = \sqrt{n}h$, they will be contained in Hellinger balls of radius $a_\nu/\sqrt{n}$ with centers at distance at least $b_\nu/\sqrt{n}$. Here we can take full Hellinger balls in the whole space of all probability measures on the σ-field of individual observations. Let them be $B^1_{\nu,i}$ and $B^1_{\nu,j}$. Let $B_{\nu,i}$ be the set of joint distributions of independent observations $X_1, X_2, \ldots, X_n$ where the distribution of X_k is any arbitrary member of $B^1_{\nu,i}$.

It follows from some considerations of Le Cam [1973] and deeper results of Birgé [1983] that there exist test functions $\varphi_{\nu,i,j} = \varphi^2_{\nu,i,j} = 1 - \varphi_{\nu,j,i}$ such that

$$\int(1 - \varphi_{\nu,j,i})dP + \int \varphi_{\nu,i,j} dQ \leq 2e^{-(b_\nu - 2a_\nu)^2}$$

for all $P \in B_{\nu,i}$ and $Q \in B_{\nu,j}$.

Thus the number $2\beta(\nu)$ of Proposition 5 is at most $2\exp\{-(b_\nu - 2a_\nu)^2\}$.

Now consider the number $C(\nu)$. It is at most equal to the number of sets of diameter a_ν needed to cover a ball of radius $a_\nu + b_{\nu-1}$.

Note that since we have taken $b_\nu = b_0 q^\nu$ and $a_\nu = b_\nu / r$, the ratio of the two diameters is

$$\frac{2(a_\nu + b_{\nu-1})}{a_\nu} = 2(1 + \frac{r}{q}).$$

This suggests a definition of dimension, related to Kolmogorov's definitions of metric entropy and capacity.

Definition 5. *A set Θ metrized by W has dimension $D(\tau)$ or less at level $\tau \geq 0$ if every subset of Θ of diameter $2x$, $x \geq \tau$, can be covered by no more than $2^{D(\tau)}$ sets of diameter x.*

Note that if any set of diameter $2x$, $x \geq \tau$ can be covered by $2^{D(\tau)}$ sets of diameter x, then any set of diameter Kx, $x \geq \tau$ can be covered by no more than $2^{mD(\tau)}$ sets of diameter x for an integer m such that $K \leq 2^m$. Thus if our set Θ metrized by W has dimension $D(a_\nu)$, every set of diameter at most $2(1+r/q)a_\nu$ can be covered by $2^{mD(a_\nu)}$ sets of diameter a_ν where m is the smallest integer that is at least equal to $\log_2 2(1 + r/q)$.

In summary, the number $C(\nu)$ of Proposition 5 will not exceed $2^{mD(a_\nu)}$. In addition, since $D(\tau)$ increases as τ decreases, $C(\nu)$ will not exceed $2^{mD(\tau)}$, where τ is the last value a_k considered in the sum in the proposition. The numbers r and q can be selected fairly arbitrarily. The choice of $r = 4$ and $q = 1/\sqrt{2}$ gives a result as follows.

Proposition 6. *Let Θ be metrized by W with $W^2(s,t) = nh^2(s, t)$. Assume that Θ has a dimension function $D(\tau)$ for $\tau \geq 0$. Then there exists confidence sets S such that $P_\theta[\theta \in S] \geq 1 - \alpha$ and their diameters do not exceed b for the smallest b that satisfies the inequality*

$$b^2 \geq 2 \log \frac{4}{\alpha} + 8D(\frac{b}{4}).$$

This follows by crude evaluation of the sums in Proposition 5 and appropriate selection of the cutoff point b_k.

Note that here the metric used is $W = \sqrt{n}h$. If written in terms of the individual distances h, the inequality would read:

diameters $c_n = b/\sqrt{n}$ that satisfy

$$c_n^2 \geq \frac{2}{n} \log \frac{4}{\alpha} + \frac{8}{n} D_h(\frac{c_n}{4})$$

where the dimension D_h is also taken for the metric h.

The corollary of Proposition 5 can also be used to bound risks such as $E_\theta W^2(\hat{\theta}_n, \theta)$. One gets upper bounds of the type

$$E_\theta W^2(\hat{\theta}_n, \theta) \leq C_1 + C_2 D$$

where D is a value of $D(\tau)$ taken at a τ that satisfies approximately $\tau^2 = C_3 D(\tau)$ for certain universal constants C_i, $i = 1, 2, 3$. For this and other results see Birgé [1983] and Le Cam [1986].

Le Cam [1975, 1986] has also extended these results to the case of observations that are independent but not necessarily identically distributed. The extension depends heavily on inequalities proved by Birgé [1984]. The argument is considerably more complex; see Le Cam [1986, p. 483–492].

Note that these inequalities involve the dimension function $\tau \rightsquigarrow D(\tau)$. They are particularly simple if $\sup_\tau D(\tau) < \infty$. However, they remain applicable even if $D(\tau) \to \infty$ as $\tau \to 0$. In such a case they can be used to obtain feasible rates of convergence for estimates. Since the dimension concept is clearly related to the "dimension" of Assouad's hypercubes (see Lemma 12), one is led to wonder whether the dimension function D is a crucial entity in estimation problems. It is easy to construct examples where $D(\tau) = \infty$ for all interesting τ but where there exist estimates $\hat{\theta}_n$ with $\sup_\theta E_\theta W^2(\hat{\theta}_n, \theta)$ growing at arbitrarily slow rates. However, except for the case where $E_\theta W^2(\hat{\theta}_n, \theta)$ stays bounded independently of θ and n, these examples cannot be considered "natural." Indeed, Birgé [1984] has shown that if the rates of convergence are not dimension controlled, then small perturbation of the measures p_θ will lead to the impossibility of achieving the same rates of convergence. For these small perturbations the rates will be dimension controlled.

That there are many examples where the results of Proposition 6 or similar ones are applicable can be seen from the papers of Birgé [1983]. See also Yatracos [1985].

Note that, in Proposition 6, the distance or loss function used is $W(s,t) = \sqrt{n}h(s,t)$. This means that the estimation problem is that of estimating the entire measure p_θ, not simply a function or functional of it. As mentioned before, rates of convergence for various functionals have been obtained by Donoho and Liu [1991] and others. They can be much better than rates for the measure itself.

7.6 Some Cases Where the Number of Observations Is Random

Consider a sequence $\{X_1, X_2, \ldots, X_n, \ldots\}$ of independent observations with common individual distribution p_θ; $\theta \in \Theta$. We have given in the preceding sections some indications of what can happen if the statistician observes $\{X_1, \ldots, X_n\}$ where n is some nonrandom number that tends to infinity.

One gets a more general theory if one lets the observations themselves decide when to stop observing, thus using $\{X_1, X_2, \ldots X_N\}$ where N is a random variable. Here we shall only consider variables N that are *stopping times* of the sequences $\{X_1, X_2, \ldots\}$ in the sense that the set where $N = k$ is a function of the variables $X_1, X_2, \ldots X_k$ and perhaps of some variable uniformly distributed in $[0,1]$, independently of θ, to allow for randomization of the stopping rule.

Such schemes have been proposed by several authors, most particularly by Wald [1947] who wrote a book on sequential analysis. The schemes can be more economical in terms of number of observations than schemes with nonrandom sample size. This is particularly true when testing hypotheses. For estimation purposes, under regularity conditions, they are not all that economical. See, for instance, Chao [1967] and the last section of chapter 17 in Le Cam [1986]. Their real value in estimation problems is that they afford possibilities that are not available with nonrandom sample sizes. A prime example is the following construction due to Stein [1945]. On a sequence $\{X_1, X_2, \ldots,\}$ where the X_j have an $\mathcal{N}(\mu, \sigma^2)$ distribution one can find stopping times N that will allow the construction of confidence intervals of preassigned width and coverage independently of σ. Another example arises in the estimation of θ for a uniform distribution on $[\theta - 1/2,\ \theta + 1/2]$. One waits for a time N when

$$\max[X_1, X_2, \ldots, X_N] - \min[X_1, X_2, \ldots X_N] \geq 1 - \epsilon.$$

Then one can guarantee that the average of the minimum and maximum of $X_1, X_2, \ldots X_N$ is within ϵ of θ.

Sometimes it is not so clear what should be called "random sample size." Consider, for example, the problem of counting bacteria under a microscope. Assuming that the microscope slide carries a grating, one can either count bacteria in a prespecified number of cells of the grating or count as many of those cells as needed to see, for example, 10 bacteria. In one case, the number of bacteria is random. In the other, the number of cells counted is random, but one would not normally classify the procedure as one where "the number of observations is random."

In any event, our purpose here is not to give a theory of sequential analysis, but only to point out that such sequential schemes lead easily to situations where the LAQ or LAMN conditions are satisfied but the LAN conditions are not. In the nonrandom sample size cases studied in previous sections we do not know of natural examples satisfying LAQ but not LAN. That is true at least if one takes n i.i.d. observations from fixed measures p_θ. If one takes n i.i.d. observations from measures $p_{\theta,n}$ that depend on n, the situation is very different. Examples satisfying LAMN but not LAN abound. We already alluded to such possibilities in Section 4. However, here is a particular example.

Example 5. Take on $\Re^k$ a random matrix M that is almost surely symmetric and positive definite. Let Y be a random vector that is $\mathcal{N}(0, M)$ conditionally, given the matrix M. The distribution of the pairs (Y, M) is a certain measure, say, Q_0. Take for Q_θ; $\theta \in \Re^k$ the measure that has density $\exp\{\theta' Y - (1/2)\theta' M \theta\}$ with respect to Q_0. Then the family $\{Q_\theta; \theta \in \Re^k\}$ is LAMN, being in fact exactly mixed normal.

Now suppose that the distribution of the matrix M is also infinitely divisible. Then, for each integer n, Q_θ can be written as a product measure $Q_\theta = \prod_{j=1}^n p_{\theta,j}$, joint distribution of n independent identically distributed observations. To do this, it is sufficient to represent M as a sum $M = \sum_{j=1}^n M_{j,n}$ of independent identically distributed matrices $M_{j,n}$. The $M_{j,n}$ will automatically be positive semidefinite. Indeed positive-semidefinite matrices form a convex cone and the support of M is obtained by taking the smallest closed semigroup (under addition) that contains the support of the $M_{j,n}$. Therefore there will be independent variables $Y_{j,n}$ that, conditionally on $M_{j,n}$ have a $\mathcal{N}(0, M_{j,n})$ distributions, say, $p_{0,j}$. One can then take for $p_{\theta,j}$ the measure that has density $\exp\{\theta' Y_{j,n} - (1/2)\theta' M_{j,n}\theta\}$ with respect to $p_{0,j}$. This gives the desired product representation.

However, we digress. Let us go back to our random number of observations N, where N is a stopping time of a given infinite sequence $\{X_1, X_2, \ldots,\}$ of i.i.d. variables.

This gives a certain specific experiment. To have an asymptotic theory it will be necessary to let the random N tend to infinity in probability. Thus we will be looking again at a sequence $\{\mathcal{E}_n; n = 1, 2 \ldots\}$ of experiments, all based on the same sequence $\{X_1, X_2, \ldots,\}$ but with stopping times N_n that tend to infinity in probability.

Note that the letter "n" used here is no longer the sample size. It is there just to tend to infinity. We shall also need some sequence of numbers, say, a_n, to indicate the approximate size of the sample and relate that to the size of neighborhoods in Θ over which approximations are sought. The reader might think of a_n as a median of N_n or as any other characteristic that reflects the general size of N_n. It need not be EN_n; that expectation may be infinite. For instance, N_n might be the nth return to equilibrium in a standard random walk obtained by coin tossing. Then $EN_n = \infty$ and even $EN_n^{1/2} = \infty$. However, the median of N_n is of the order of $2n^2$. This could be taken as a_n for such stopping times.

However, in order to make formulas look like the formulas of the previous sections, we shall pretend in the sequel that $a_n = n$.

This does not actually entail any serious loss of generality. It is really a matter of notation, but it will allow us to use sequences $t/\sqrt{n}$ or $t_n/\sqrt{n}$ with $|t_n|$ bounded, as in Section 7.2. However, we shall need a very specific assumption on the behavior of N_n/n or, if one so desires, N_n/a_n. It is as follows.

Assume that $\theta \in \Theta \subset \Re^k$ and that the individual distributions $p_\theta; \theta \in \Theta$ satisfy DQM_o at $\theta = 0$. Assume in addition the following.

Assumption 2. *For the measures induced by p_θ, $\theta = 0$, the variables N_n are stopping times such that N_n/n has a relatively compact sequence of distributions.*

Under this assumption one can state a result very similar to that of Section 7.2. Using the same notation as in Section 7.2, with V_j for derivatives in quadratic mean, it is as follows.

Proposition 7. *Let the individual measures p_θ; $\theta \in \Re^k$ satisfy DQM_o at $\theta = 0$. Let the stopping times N_n satisfy Assumption 2 (at $\theta = 0$). Let $Q_{\theta,n}$ be the joint distribution of $\{N_n, X_1, X_2, \ldots, X_{N_n}\}$ and let $\Lambda_n(\theta)$ be the logarithm of likelihood ratio*

$$\Lambda_n(\theta) = \log \frac{dQ_{\theta,n}}{dQ_{0,n}}.$$

Then, for $|t_n|$ bounded, the quantity

$$\Lambda_n(\frac{t_n}{\sqrt{n}}) - \frac{1}{\sqrt{n}} t'_n \sum_{j=1}^{N_n} V_j + \frac{1}{2}\frac{N_n}{n} E(t'_n V_1)^2$$

tends to zero in $Q_{0,n}$ probability.

Remark 1. Note that the sequences $t_n/\sqrt{n}$ are nonrandom and that we have not replaced them by items such as $t_n/\sqrt{N_n}$. Note also that the convergence in probability is under $Q_{0,n}$ only. It is not claimed here that the LAQ conditions hold.

Proof. The argument is almost the same as that of Proposition 1, Section 7.2. One takes the square roots of likelihood ratios called $Z_j(\theta + t_n/\sqrt{n})$ there and writes remainder terms

$$Z_j(\theta + \frac{t_n}{\sqrt{n}}) - 1 - \frac{1}{2\sqrt{n}} t'_n V_j + h^2(\theta + \frac{t_n}{\sqrt{n}}, \theta) = \frac{|t_n|}{\sqrt{n}} R_j(\theta, \frac{t_n}{\sqrt{n}}).$$

The compactness condition of Assumption 2 implies that for every $\epsilon > 0$ there is a number b such that $Q_{0,n}[N_n > bn] < \epsilon$. Now look at the sum

$$n^{-1/2} \sum_j \{R_j(\theta, \frac{t_n}{\sqrt{n}}); \ 1 \leq j \leq N_n \wedge (bn)\}.$$

Conditionally given the variables $X_1, X_2, \ldots, X_{j-1}$, the term R_j has expectation zero and a variance σ_n^2 that tends to zero as $n \to \infty$. Thus the sum up to $N_n \wedge (bn)$ tends to zero in quadratic mean. It follows that

$$n^{-1/2} \sum_j \{R_j(\theta, \frac{t_n}{\sqrt{n}}); \ 1 \leq j \leq N_n\}$$

also tends to zero in probability.

The passage from sums of square roots to sums of logarithms can be carried out as usual; see Lemma 3, Section 5.3. The result follows from the easily verifiable fact that $\Lambda_n(t_n/\sqrt{n})$ is indeed equal to the sum up to N_n of the logarithms of likelihood ratios of the individual components. □

We already noted that Proposition 7 does not say that the LAQ conditions are satisfied. This is because the convergence statement is under $Q_{0,n}$, not necessarily under $Q_{t_n/\sqrt{n},n}$. Thus, we shall add another restriction as follows.

Assumption 3. *If $\{|t_n|\}$ stays bounded, then the sequence of distributions $\mathcal{L}\left[N_n/n | Q_{t_n/\sqrt{n},n}\right]$ is relatively compact.*

Simple examples show that Assumption 3 is not a consequence of DQM_o and Assumption 2.

This leads to the following corollary of Proposition 7.

Corollary 3. *Let DQM_o be satisfied at $\theta = 0$ and let Assumption 3 be satisfied. Then the measures $Q_{\theta,n}$ will satisfy at $\theta = 0$*

the LAQ conditions of Chapter 6. If the matrix EV_1V_1' does not vanish, they will satisfy the LAN conditions if and only if the N_n/n have a limiting distribution degenerate at a point.

Proof. Under Assumption 3, for each $\epsilon > 0$ there will exist finite numbers b such that $Q_{0,n}[N_n > bn] < \epsilon$ and $Q_{t_n/\sqrt{n},n}[N_n > bn] < \epsilon$.

The product measures distributions of $X_1, X_2, \ldots, X_{bn}$ are contiguous, one can show that the same will be true of the less informative measures, distributions of $X_1, X_2, \ldots, X_{N_n \wedge (bn)}$. The argument can be found in Yang [1968] and Le Cam and Yang [1988]. Since ϵ is arbitrary the conclusion extends to the measures $Q_{0,n}$ and $Q_{t_n/\sqrt{n},n}$ themselves. The statement about the LAN conditions follows from the fact that, in the LAN case, the homogeneous quadratic term must be nonrandom. □

Here we are in the standard i.i.d. case where the measures p_θ are defined on a fixed space and do not vary with n. It then follows from the mixing theorems of Rényi [1958] (see also Wittenberg [1964]) that if N_n/n tends in probability to some random variable $\xi > 0$, then the LAMN assumptions of Chapter 6 will be satisfied.

However, this requires convergence *in probability,* not only convergence *in distribution.* (Such convergence in probability to a random ξ does not make much sense if, as recommended, one would let p_θ and the space on which it is defined depend on n.)

The situation becomes clearer if one uses some heavy artillery in the form of "invariance principles." Those have nothing to do with invariance under group transformations. They are the accepted name for some functional central limit theorems.

To give an inkling of what happens, let θ be one-dimensional. That is, suppose $\theta \in \Re$ instead of $\Re^k$.

Introduce a "time" τ as $\tau(j) = j/n$ if j observations have been taken. Under p_0, the sum $n^{-1/2}\sum_{i=1}^{j} V_i$ of Proposition 7 will then tend to a Wiener process $\{W(\tau); \tau \in [0,\infty)\}$ that is to a gaussian process with mean zero and variance $E[W(\tau)]^2 = \sigma^2\tau$ where we can suppose that $\sigma^2 = 1$ for simplicity. The stopping variable N_n/n becomes a stopping time τ_n of that process with the proviso of Assumption 2 that, under $\theta = 0$, τ_n has a relatively compact sequence of distributions (under p_0.) The log likelihoods would become

$$tW(\tau_n) - \frac{1}{2}t^2\tau_n$$

for deviations $\theta_n = t/\sqrt{n}$ from $\theta_0 = 0$, showing that under the alternatives $\theta + (t_n/\sqrt{n})$ the Wiener process $\{W(\tau); \tau \geq 0\}$ acquires a drift equal to τt_n.

On such a Wiener process, it is clear that, given τ_n the distribution of $W(\tau_n)$ will not necessarily be gaussian. In fact, since one can select τ_n to give $W(\tau_n)$ any distribution one pleases, including distributions with bounded support, provided that $EW(\tau_n) = 0$ if one wants $E\tau_n < \infty$, there will be many cases where LAQ will be satisfied at $\theta = 0$, but LAMN will not.

The passage to Wiener processes described above can be justified. See, for example, Le Cam [1979a] for the part about distances between experiments and any text about convergence of stochastic processes for the weak convergence to Wiener processes. Contrary to generally received opinion, these did not originate with Donsker [1951] but, at least, with Kolmogorov [1931, 1933].

This does not prevent the fact that, if we require DQM_o and Assumption 3 not only at $\theta = 0$ but at all $\theta \in \Re^k$, the LAMN conditions will be satisfied almost everywhere (see Section 5.5)

7.7 Historical Remarks

The statistical literature on asymptotics for the independent identically distributed case is absolutely huge. It can be said to have started with Laplace [1810a, b]. Some of his contributions will be discussed further in Chapter 8. The remainder of the 19th century does not seem to have produced many lasting contributions, but we have not looked at the matter very carefully and may have missed some.

A noteworthy effort appears in Edgeworth [1908, 1909]. He attempted to prove that estimates "having the property of an average" cannot asymptotically behave any better than estimates obtained by maximizing posterior densities. The proofs given by Laplace or Edgeworth cannot be called rigorous. Their statements depend on assumptions that are not clearly stated. For a clear statement along Edgeworth's lines, see Wilks [1938].

The subject gained considerable momentum with the publication of Fisher's papers [1922, 1925]. Fisher's technique is akin to that of Laplace or Edgeworth, but he recognized certain central ideas and gave attractive names to many concepts such as "consistency," "efficiency," "sufficiency," "likelihood," and so on. Fisher's statements, although perhaps not entirely rigorous, had a profound impact. Among attempts to do something that would be more satisfactory in a mathematical sense, one can mention two papers of Doob [1934, 1936] and one by Wilks [1938]. There was also closely related work of Dugué [1936a, 1936b, 1937].

The war came along, causing much disruption, but also bringing Wald to the United States. Under the influence of Hotelling, Wald turned his considerable skills toward statistics. He generated a flood of new ideas, treated with enormous analytical power. His 1943 paper represents a turning point. He shows that whatever one can do with the actual distributions of the observations, one can achieve nearly as well working with the normal approximations to the distributions of maximum likelihood estimates. Similar ideas were later used by Le Cam [1956] and other authors. They form the basis of the present notes.

Most of that work was done under variations of what are called "Cramér's conditions." The name comes from an account given by Harald Cramér in his 1946 treatise. Note that in Stockholm Cramér did not have access to Wald's work of 1943. The war made communication difficult; that is reflected in several duplications of effort. An example in point is the Cramér-Rao inequality simultaneously discovered by Cramér in Sweden and Rao in India, but actually proved earlier in papers by Fréchet [1943] and quickly extended to the multivariate situation by Darmois [1945].

The conditions of the Cramér type always involved first and

second derivatives sometimes three derivatives of the likelihood function. Daniels [1961] used only first derivatives but with other restrictions. That many of the results remain valid with only a first derivative, but in quadratic mean, was duly noted in conversations between Hájek and Le Cam in 1962. The first publication using derivatives in quadratic mean seems to be Hájek [1962]. He gives a version of Lemma 3, Section 7.3. Actually the passage from two derivatives to only one involved another decision: to go away from maximum likelihood estimates themselves and use instead techniques proposed in connection with the LAN assumptions of Le Cam [1960].

The examples of Section 7.3 are classical by now. For the densities that are exponential in a power of $|x|$, credit must be given to Daniels [1961] and Prakasa Rao [1968].

Some of the facts used in Section 7.2, for instance, the differentiability almost everywhere of a function satisfying $\limsup_{s\to t} h(s,t)/|s-t| < \infty$ at each t can be easily derived from Gelfand [1938] or Alexiewicz [1950] when the parameter is real. Le Cam [1970] uses a method of Saks [1937] to extend the result to euclidean spaces. The method, or a similar one, is said to have been used by Rademacher much earlier.

The introduction of the local Hilbert space called $\mathcal{L}_{2,0}(p)$ here seems to have occurred simultaneously from several quarters. One can mention Levit [1973] and Moussatat [1976]. It is now very widespread.

Note, However, that the use of some Hilbert space is implicit in the use of the Hellinger distance or a multiple such as $\sqrt{n}h$. In the independent identically distributed case, the Hilbert space happens to have a nice concrete form. For independent variables that are *not* identically distributed, the use of the square distance $\sum_j h_j^2$ still implies the use of Hilbert spaces. They just do not have the simple form of the i.i.d. situation.

What we have called "gaussian auras" are described in Le Cam [1985]. The results stated here without proof are proved in detail there. We have mentioned them here because there are still open problems and because we wanted to emphasize that, when n tends to infinity, one needs to look at varying sets B_n

of parameter values and should not restrict oneself to fixed sets, B, independent of n.

The description of a problem of estimation for linear functions is, of course, very simple in the full gaussian model.

That one could do something similar in the asymptotic of the i.i.d. case was noted by Stein [1956], Beran [1974], Stone [1975] and others. It has now become a full-fledged industry called "semiparametrics." A volume on the subject, authored by Bickel, Klaassen, Ritov and Wellner, appeared in 1993.

There is a sizeable literature on lower bounds for risks of estimates. The Cramér-Rao bound was proposed independently by Cramér [1946] and Rao [1945]. It had been obtained earlier by Fréchet [1943] and Darmois [1945]. In our account it follows from considerations that involve only a pair of values of the parameter points. For a formulation that uses weak derivatives instead of strong ones, see Fabian and Hannan [1977]. It should be mentioned that Barankin [1949] gave a very general formulation using the Riesz, or Hahn-Banach evaluation of norms for linear functionals.

We have taken Fano's lemma from the treatise by Ibragimov and Has'minskii [1981]. The original lemma occurs in R. M. Fano's unpublished notes [1954]. Assouad's lemma appears in Assouad [1983]. It codifies a procedure used by Bretagnolle and Huber [1979] and Birgé [1983].

Upper bounds for risks of estimates have usually been obtained by producing particular estimates. Between 1950 and 1955 several authors looked at the problem of consistent estimation in the i.i.d. case. One early result was given by Stein: there is no consistent test of whether θ is rational or irrational for the $\mathcal{N}(\theta, 1)$ distributions. Hoeffding and Wolfowitz [1958] gave various sufficient conditions in a sequential setup. Le Cam and Schwartz [1960] gave necessary and sufficient conditions, always for the i.i.d. case. The fact that these conditions involved uniform structures instead of simpler topological ones was somewhat of a surprise.

The appeal to dimensionality restrictions with an analogue of Kolmogorov's definition of metric entropy and metric capacity (see Kolmogorov and Tichomirov [1959]) appears in Le Cam [1973]. Because of some difficulties with the statements in that paper, it was redone in Le Cam [1975]. Still, lemma 1 of Le Cam [1975] is obviously erroneous. Its statement and proof lack a number of summation signs. The final results remain valid, perhaps with a modification of constants; see Le Cam [1986, chapter 5, section 5, p. 78].

Le Cam [1975] had given an extension to independent observations that are not necessarily identically distributed. It was considerably improved by the use of a clever argument of Birgé [1983]. Le Cam [1975] used a successive refinements approach as described here in Section 5. Birgé [1983] proposed a different technique, covering the entire space by Hellinger-type balls of the same but well-selected radius. For more indications on the role of the metric dimension, see Birgé [1983, 1984]. Most of this applies to the case where one wants to estimate the entire underlying distributions, as in the estimation of densities.

The situation for estimation of real-valued functions defined on the space of measures is rather different; see Donoho and Liu [1991].

The results on random numbers of observations given in Section 7.6 were intended only for illustration. Many more illustrations can be obtained from stochastic processes. A proper treatment of the subject would involve quoting the works of Aalen [1978], Jeganathan [1983], Liptser and Shiryaev [1977, 1978], Shiryaev [1981], Greenwood and Shiryayev [1985], and many other authors.

8

On Bayes Procedures

8.1 Introduction

In this chapter we describe some of the asymptotic properties of Bayes procedures. These are obtained by using on the parameter set Θ a finite positive measure μ and minimizing the average risk $\int R(\theta, \rho)\mu(d\theta)$. (See Chapter 2 for notation.) The procedure ρ that achieves this minimum will, of course, depend on the choice of μ. However, the literature contains numerous statements to the effect that, for large samples, the choice of μ matters little. This cannot be generally true, but we start with a proposition to this effect. If instead of μ one uses λ dominated by μ and if the density $d\lambda/d\mu$ can be closely estimated, then a procedure that is nearly Bayes for μ is also nearly Bayes for λ.

Then we recall a result of Doob [1948] that says that under weak conditions, Bayes procedures are consistent almost surely almost everywhere. Even better, they tend to converge at the "right" rate. These results will be given in Section 8.2. In Section 8.3, we will describe a form of what is called the Bernstein–von Mises theorem even though it was already noted by Laplace [1810a, b]. We have not attempted to give the most general available form of the result, but we point out the various steps needed for a proof. Section 8.4 gives a set of sufficient conditions for the Bernstein–von Mises theorem in the i.i.d. case.

Section 8.5 recalls, without proof, several results mostly due to Freedman and Diaconis, to the general effect that due caution must be exerted. Sets of measure zero for the prior μ may be nearly the entire space in a topological sense and, in that sense, Bayes procedures can misbehave nearly always.

8.2 Bayes Procedures Behave Nicely

Consider a sequence $\{\mathcal{E}_n\}$ of experiments $\mathcal{E}_n = \{P_{\theta,n} : \theta \in \Theta\}$ given by measures on σ-field $\mathcal{A}_n$. Let us assume also that for each n one is given a set D_n of possible decisions and a loss function W_n on the product $D_n \times \Theta$. We shall assume here that these objects satisfy the following requirements.

(1) There is on Θ a σ-field $\mathcal{B}$ such that all functions $\theta \rightsquigarrow P_{\theta,n}(A)$, $A \in \mathcal{A}_n$ are $\mathcal{B}$-measurable.

(2) The decision spaces D_n carry σ-fields $\mathcal{D}_n$ such that the loss functions W_n are $\mathcal{D}_n \times \mathcal{B}$-measurable.

(3) The loss function W_n is such that $0 \leq W_n(z, \theta) \leq 1$ for all $z \in D_n$ and $\theta \in \Theta$.

(4) One has chosen on $(\Theta, \mathcal{B})$ a particular finite positive measure μ.

In this situation one can define various numbers as follows. Let $R_n(\theta, \rho)$ be the risk at θ of a decision procedure ρ. Let λ be any positive finite measure on $\mathcal{B}$. Write $R_n(\lambda, \rho)$ for the integrated risk $\int R_n(\theta, \rho)\lambda(d\theta)$. Let $\chi(\lambda)$ be the Bayes risk $\chi(\lambda) = \inf_\rho R_n(\lambda, \rho)$ and let $K_n(\lambda, \rho) = R_n(\lambda, \rho) - \chi(\lambda)$ be the "regret" of ρ at λ.

If φ is a real-valued function defined on Θ and $\mathcal{B}$-measurable there, let

$$\alpha_n(\varphi) = \inf_h \int\int |\varphi(\theta) - h(x)| P_{\theta,n}(dx)\mu(d\theta)$$

for an infimum taken over all real-valued $\mathcal{A}_n$-measurable functions that are integrable for the measure $P_{\theta,n}(dx)\mu(d\theta)$.

Proposition 1. *Under conditions (1) to (4), the regret functions K_n satisfy the following inequality. Let λ be the measure given by $d\lambda = \varphi d\mu$ and let $d\lambda' = (1 - \varphi)d\mu$ for a function φ that is $\mathcal{B}$-measurable and satisfies $0 \leq \varphi \leq 1$.*

Then

$$K_n(\mu, \rho) \leq K_n(\lambda, \rho) + K_n(\lambda', \rho) \leq K_n(\mu, \rho) + \alpha_n(\varphi).$$

Proof. For simplicity of notation we shall drop the subscript n. It does not play any role in the argument. Consider besides ρ

two other procedures ρ' and ρ'' and an $\mathcal{A}$-measurable function f with $0 \le f \le 1$. Let ρ''' be the procedure $\rho''' = f\rho' + (1-f)\rho''$. Define a function V on $\mathcal{X} \times \Theta$ by $V(x,\theta) = \int W(z,\theta)\rho_x(dz)$. Let V', V'' and V''' be defined similarly for the procedures ρ', ρ'' and ρ'''. Then $V''' = V' + (1-f)V''$. Since $\mu = \lambda + \lambda'$ this gives the relation

$$\begin{aligned} K(\mu,\rho) &\ge R(\mu,\rho) - R(\mu,\rho''') \\ &= [R(\lambda,\rho) - R(\lambda,\rho')] + R(\lambda',\rho) - R(\lambda',\rho'') \\ &\quad + \int [(1-f)\varphi - f(1-\varphi)][V' - V'']dS \end{aligned}$$

where $S(dx,d\theta) = P_{\theta,n}(dx)\mu(d\theta)$. Since $0 \le W \le 1$, one has $|V' - V''| \le 1$. Keeping f fixed and varying ρ' and ρ'' gives the

$$K(\mu,\rho) \ge K(\lambda,\rho) + K(\lambda',\rho) - \int |(1-f)\varphi - f(1-\varphi)|dS.$$

However, $(1-f)\varphi - f(1-\varphi) = \varphi - f$ and the right inequality follows by taking an infimum with respect to f. The other inequality follows from $\chi(\mu) \ge \chi(\lambda) + \chi(\lambda')$. Hence the results. □

This proposition says that if ρ is close to being Bayes for μ it will also be close to Bayes for λ and λ', provided that the density $d\lambda/d\mu$ can be closely estimated. The result can be applied, for instance, by taking for φ the indicator of a set $B \subset \Theta$. If the problem is to estimate θ itself, a prior measure λ carried by B will almost surely lead to an estimated value in B. If φ can be closely estimated, a Bayes estimate for μ will often know to estimate θ in B if that is where θ actually is.

To apply that to our sequence $\{\mathcal{E}_n\}$ call "accessible" any function φ defined on Θ such that $\alpha_n(\varphi) \to 0$ as $n \to \infty$.

Lemma 1. *There is a σ-field $\mathcal{B}^*$ such that all $\mathcal{B}^*$-measurable μ-integrable functions are accessible.*

Proof. If φ_1 and φ_2 are accessible, so are linear combinations $a\varphi_1 + b\varphi_2$ and so are $\varphi_1 \vee \varphi_2$ and $\varphi_1 \wedge \varphi_2$. In addition $\alpha_n(\varphi) \le \int |\varphi| d\mu$, so the space of accessible functions is closed for the convergence in the first mean for μ. Hence the result by standard arguments. □

It will often happen that $\mathcal{B}^*$ is in fact the entire σ-field $\mathcal{B}$ or its completion for μ. This follows, for instance, from a result of Doob, which we can paraphrase as follows. Suppose that the σ-fields $\mathcal{A}_n$ are all on the same space, say, $\mathcal{X}$. Suppose also that $\mathcal{A}_n \subset \mathcal{A}_{n+1}$ for all n. Let $\mathcal{A}_\infty$ be the σ-field generated by $\cup_n \mathcal{A}_n$.

Proposition 2. *With the conditions just described suppose that $P_{\theta,n}$ is the restriction to $\mathcal{A}_n$ of a measure $P_{\theta,\infty}$ defined on $\mathcal{A}_\infty$. Assume also that there is a -measurable function f from $(\mathcal{X}, \mathcal{A}_\infty)$ to $(\Theta, \mathcal{B})$ such that $\int\int |\theta - f(x)| P_{\theta,\infty}(dx)\mu(d\theta) = 0$. Then all $\mathcal{B}$-measurable μ-integrable functions φ are accessible.*

The proof will be given after Proposition 3.

For Doob's result, assume that $(\Theta, \mathcal{B})$ is a Borel subset of a complete separable metric space with its σ-field of Borel subsets. Disintegrate the measures $P_{\theta,n}(dx)\mu(d\theta)$ in the form $F_{x,n}$ $(d\theta)S'_n(dx)$. Call the $F_{x,n}$ consistent at θ if for every neighborhood V of θ the posterior measure $F_{x,n}(V^c)$ tends to zero almost surely for $P_{\theta,\infty}$.

Proposition 3. (Doob [1948].) *Under the conditions of Proposition 2 and with the additional assumption that $(\Theta, \mathcal{B})$ is a Borel set in a complete separable metric space, as described, for μ almost all $\theta \in \Theta$, the posterior measures $F_{x,n}$ are consistent.*

Proof. Both propositions follow from Doob's martingale convergence theorem. To see this assume that μ is a probability measure and take an arbitrary $\mathcal{B}$-measurable bounded function φ on Θ. By the assumption made in Proposition 2, one can write $\varphi(\theta) = \varphi[f(x)]$ almost surely for the joint measure $S(dx, d\theta) = P_{\theta,\infty}(dx)\mu(d\theta)$ on $\mathcal{X} \times \Theta$. Then $\int \varphi(\theta) F_{x,n}(d\theta)$ is a version of the conditional expectation $E[\varphi|\mathcal{A}_n]$ of $x \rightsquigarrow \varphi[f(x)]$. By Doob's martingale convergence theorem, $E[\varphi|\mathcal{A}_n]$ converges almost surely to $E[\varphi|\mathcal{A}_\infty]$. By Lemma 1 this gives Proposition 2. To obtain Proposition 3 one can note that $F_{x,n}(V^c) \to 0$ for every neighborhood V of θ is equivalent to the ordinary weak convergence of measures to the mass δ_θ at θ. Here, under the assumption that Θ is Borel in a complete separable metric space, this is equivalent to convergence of $\int \varphi(\tau) F_{x,n}(d\tau)$ to $\varphi(\theta)$ for a certain *countable* family of bounded measurable functions φ. Thus, by Proposition 2, the set in $\mathcal{X} \times \Theta$ where $F_{x,n} - \delta_\theta$ tends to zero has measure unity for S. Its complement N has measure zero. Hence, by Fubini's theorem almost all the sections of N for constant θ are sets of measure zero. Hence the result. □

Doob's [1948] result contained another important fact. He considers i.i.d. observations with common distributions $\{p_\theta; \theta \in \Theta\}$ where the p_θ are on a Euclidean space and $p_s \neq p_t$ if $s \neq t$. He shows then that the existence of the function f of Proposition 2 is assured whenever $(\Theta, \mathcal{B})$ is Borelian as in Proposition 3. Here this means that for appropriate structures on $(\mathcal{X}, \mathcal{A}_\infty)$ the fact that measures $P_{s,\infty}$ and $P_{t,\infty}$, $s \neq t$ are two by two disjoint already implies the existence of f with $f(x) = \theta$ almost surely. We shall not elaborate further on this point here.

To describe still another good property, let us keep the assumption that $(\Theta, \mathcal{B})$ is a Borel subset of a complete separable metric space, but return to the situation where the σ-fields $\mathcal{A}_n$ are arbitrary on spaces $\mathcal{X}_n$ that can vary with n.

Instead of using a fixed prior measure μ, let us take arbitrary positive finite measures μ_n. Define a joint distribution S_n by $S_n(dx, d\theta) = P_{\theta,n}(dx)\mu_n(d\theta) = F_{x,n}(d\theta)S'_n(dx)$. Let us say that estimates T_n converge at a rate δ_n, $\delta_n \to 0$, if for every $\epsilon > 0$ there is a $b \in (0, \infty)$ such that

$$\limsup_n S_n\{d(T_n, \theta) > b\delta_n\} < \epsilon,$$

where $d(T_n, \theta)$ is the distance between T_n and θ in the metric space Θ.

Now for each x and n, define a number $\gamma_0(x, n)$ as the infimum of the numbers γ for which there is in Θ a ball B of radius γ such that $F_{x,n}(B) > 1/2$. If $\gamma_0(x, n) > 0$, let $\gamma_1(x, n) = (1 + n^{-1})\gamma_0(x, n)$. If $\gamma_0(x, n) = 0$, let $\gamma_1(x, n) = 2^{-n}\delta_n$. Then, by definition, there will be a ball $B(x, n)$ of radius $\leq \gamma_1(x, n)$ such that $F_{x,n}[B(x, n)] > 1/2$.

Take for $\tilde{\theta}_n(x)$ any point in $B(x, n)$.

Proposition 4. *Assume that $(\Theta, \mathcal{B})$ is Borelian as stated and construct estimates $\tilde{\theta}_n$ as described. Then, if there are estimates that converge at the rate δ_n, so will the $\tilde{\theta}_n$.*

Note 1. For this to make good sense we should select the $\tilde{\theta}_n$ in a measurable way; this can be done. Alternatively, one can state that the outer measures $S_n^*[d(\tilde{\theta}_n, \theta) \geq b\delta_n]$ will eventually be less than ϵ.

Proof. Take an $\epsilon > 0$ such that $\epsilon < 1/4$ and let $D(x, n)$ be the set $D(x, n) = \{\theta : d(T_n(x), \theta) < b\delta_n\}$, for a number b such that $S_n\{d(T_n, \theta) \geq b\delta_n\} < \epsilon$. This inequality can also be written $\int F_{x,n}[D^c(x, n)]S_n'(dx) < \epsilon$. Now let A_n be the set of values of x such that $F_{x,n}[D^c(x, n)] \geq 1/2$. By Markov's inequality $S_n'(A_n) < 2\epsilon$.

If x does not belong to A_n then both $D(x, n)$ and $B(x, n)$ have probability $> 1/2$ for $F_{x,n}$. Thus they must intersect. Therefore one must have

$$d[T_n(x), \tilde{\theta}_n(x)] \leq 2[\gamma_1(x, n) + b\delta_n].$$

However, $D(x, n)$ is a ball with radius $b\delta_n$ that has probability $> 1/2$. Thus $\gamma_0(x, n) \leq b\delta_n$.

Thus $2[\gamma_1(x, n) + b\delta_n] \leq (1 + n^{-1})4b\delta_n$ unless $\gamma_0(x, n) = 0$ in which case

$$2[\gamma_1(x, n) + b\delta_n] \leq 2\delta_n[b + 2^{-n}].$$

The result follows. □

Note that this result involves two separate rates of convergence if one takes measures $\mu_n = c_n\nu_n$ for numbers $c_n > 0$ and probability measures ν_n. There is the rate δ_n at which the balls shrink and the rate c_n^{-1} at which the tail probabilities go to zero.

All of this seems to imply that Bayes procedures are generally well behaved.

However, note that the results involve mostly the joint measures S_n or for fixed μ, sets of μ-measure zero. Freedman [1963] has shown that this can sometimes be poor consolation. Some of his results will be recorded in Section 5 without proofs.

8.3 The Bernstein–von Mises Phenomenon

The phenomenon has to do with the fact that, often, posterior distributions tend asymptotically to look like normal distributions. The phenomenon was noted, in the i.i.d. case, by Laplace [1810a, b]. It was further studied by Bernstein [1917] and von Mises [1931] and then by Le Cam [1953]. Since then, many extensions have been obtained.

Here we shall describe only a particular case. For simplicity, we shall assume that Θ is the entire space $\Re^k$ and that one has a sequence $\{\mathcal{E}_n\}$ of experiments $\mathcal{E}_n = \{P_{\theta,n}; \theta \in \Theta\}$. The prior measure μ will be kept fixed.

Passage to measurable subsets of $\Re^k$ and to variable prior measures μ_n can be carried out. The general principles of proof will remain the same.

Also, as in Chapter 6, we shall assume we are given a sequence $\{\delta_n\}$ of numbers $\delta_n > 0$ such that $\delta_n \to 0$. Extensions to more general cases are feasible.

It will always be assumed that the $P_{\theta,n}$ are measurable in the sense that all functions $\theta \rightsquigarrow P_{\theta,n}(A)$, $A \in \mathcal{A}_n$ are measurable. This can mean Borel measurable. One can also extend it to mean measurability for the completion of the Borel field for μ.

Under such measurability conditions one can disintegrate the joint measure $S_n(dx, d\theta) = P_{\theta,n}(dx)\mu(d\theta) = F_{x,n}(d,\theta)S'_n(dx)$.

We shall consider only convergences where there are gaussian measures $G_{x,n}$ such that $\int \|F_{x,n} - G_{x,n}\| S'_n(dx) \to 0$ or $\int \|F_{x,n} - G_{x,n}\| P_{\tau,n}(dx) \to 0$ for certain parameter values τ. Here the norm $\|F_{x,n} - G_{x,n}\|$ will be the L_1-norm (total variation).

It is perhaps more usual to state theorems saying that $\|F_{x,n} - G_{x,n}\|$ tends to zero almost surely $P_{\tau,n}$, but that does not make sense if the $P_{\theta,n}$ are on spaces $(\mathcal{X}_n, \mathcal{A}_n)$ that vary arbitrarily as n changes. Thus we shall not study this form of convergence.

First let us state a particular theorem to be able to describe what is involved.

Proposition 5. *Let $\Theta = \Re^k$ and let $\delta_n > 0$ tend to zero as $n \to \infty$. Let μ be a fixed probability measure on Θ such that μ has density f with respect to the Lebesgue measure λ on Θ.*

Assume in addition the following:

(1) *There is an $a > 0$ such that*

$$\lim_{\epsilon \to 0} \frac{1}{\lambda[B(\epsilon)]} \int_{B(\epsilon)} |f(t) - a| \lambda(dt) \to 0$$

as $\epsilon \to 0$. (Here $B(\epsilon)$ is the ball of radius ϵ centered at $\theta = 0$ in Θ.)

(2) *At $\theta = 0$ the $\{P_{\theta,n}; \theta \in \Theta\}$ satisfy the* LAQ *conditions of Chapter 6 at the rate δ_n.*

(3) *For each $\epsilon > 0$ there is a $b \in (0, \infty)$ such that if $C_n = \{\theta : |\theta| \geq b\delta_n\}$ then*

$$\limsup_n \int F_{x,n}(C_n) P_{0,n}(dx) < \epsilon.$$

Then there are gaussian measures $G_{x,n}$ such that $\int \|F_{x,n} - G_{x,n}\| P_{0,n}(dx)$ tends to zero as $n \to \infty$.

Proof. Take a set $A \in \mathcal{A}_n$ and a $B \in \mathcal{B}$. The defining equality for the joint measures shows that

$$\int_B P_{\theta,n}(A)\mu(d\theta) = \int_A F_{x,n}(B) S'_n(dx).$$

Thus if one lets M_B be the measure $M_B = \int_B P_{\theta,n}\mu(d\theta)$, it appears that $x \rightsquigarrow F_{x,n}(B)$ is on $\mathcal{A}_n$ a version of the Radon-Nikodym density dM_B/dM_Θ. Now let μ'_n be μ restricted to $B_n = C_n^c = \{\theta : |\theta| < b\delta_n\}$ and renormalized.

Let $F'_{x,n}$ be the corresponding conditional distribution given $\mathcal{A}_n$. Using the Radon-Nikodym formulas, it is easy to see that for any set $D \in \mathcal{B}$ one has

$$F'_{x,n}(D) - F_{x,n}(D) = F'_{x,n}(D) F_{x,n}(C_n).$$

Thus, since one can choose b so that $\int F_{x,n}(C_n) P_{0,n}(dx)$ is eventually less than ϵ, it will be enough to consider the behavior of $F'_{x,n}$ obtained from the measure μ'_n defined by $\mu'_n(D) = \mu(D \cap B_n)[\mu(B_n)]^{-1}$.

For these measures an argument entirely similar to the argument of Section 6.4, Proposition 4, will give the desired result. □

It should be noted that the condition that $\int F_{x,n}(C_n)P_{0,n}(dx)$ will eventually be less than ϵ is a consistency requirement at the rate δ_n for the posterior distributions $F_{x,n}$ and for $\theta = 0$. According to the results of Section 8.2 this should be true in most situations. However, as we shall see in the next section, it may take quite a bit of work to prove it.

Note 2. We have not specified the exact form of the gaussian $G_{x,n}$. It can be taken as in Section 6.4, Proposition 4, with the matrix called Γ_n there put equal to zero.

8.4 A Bernstein–von Mises Result for the i.i.d. Case

The purpose of this section is to show that the conditions of Proposition 5, Section 8.3 can actually be satisfied, but that the consistency condition of that proposition can take some effort.

Consider the following set of assumptions, valid at a particular θ_0 that we shall take $\theta_0 = 0$ for simplicity.

(A1) The set Θ is all of $\Re^k$. The measures $P_{\theta,n}$ are the joint distributions of n independent observations with common individual distribution p_θ on a σ-field $\mathcal{A}$.

(A2) If $s \neq t$ then $p_s \neq p_t$.

(A3) There is a compact $K_0 \subset \Theta$, an $\epsilon_0 \in (0, 1/2)$, and an integer N such that there is a test function ω_N based on N observations such that $\int(1 - \omega_N)dP_{0,n} < \epsilon_0$ and $\int \omega_N dP_{\tau,n} \leq \epsilon_0$ for all $\tau \in K_0^c$.

(A4) For each $A \in \mathcal{A}$ the maps $\theta \leadsto p_\theta(A)$ are Lebesgue measurable. On the compact K_0, they are continuous.

(A5) The condition DQM_o of Section 7.2, is satisfied at $\theta = 0$ and the derivative in quadratic mean V has a nonsingular covariance matrix $\Gamma = E_0(VV')$.

(A6) The prior measure μ has with respect to the Lebesgue measure λ a density f. There is a number $a > 0$ such that, $B(\epsilon)$ being the ball of radius ϵ centered at $\theta_0 = 0$, one has

$$\lim_{\epsilon \to 0} \frac{1}{\lambda[B(\epsilon)]} \int_{B(\epsilon)} |f(t) - a| \lambda(dt) = 0.$$

Under these conditions, one can prove the existence of estimates θ_n^* such that for every $\epsilon > 0$ there is a b and an N for which $P_{0,n}\{|\theta_n^*| > b\delta_n\} < \epsilon$ for all $n \geq N$ and for $\delta_n = 1/\sqrt{n}$. This will be a consequence of the proof given later. One can then construct centers Z_n by the method of Section 6.3. What they will do at values $\theta \neq 0$ is not stated here. At $\theta = 0$, they will be asymptotically normal, and the log likelihood $\Lambda_n(t, 0) = \log(dP_{t,n}/dP_{0,n})$ will have approximations of the form

$$\Lambda_n(t_n, 0) \equiv -\frac{n}{2}\{(t_n - Z_n)'\Gamma(t_n - Z_n) - Z_n'\Gamma Z_n\}$$

for all $\{t_n\}$ such that $\sqrt{n}|t_n|$ stays bounded. The matrix Γ is defined by $v'\Gamma v = E_0(v'V)^2$ for the derivative in quadratic mean of DQM_o.

Let then $G_{x,n}$ be the gaussian probability measure proportional to

$$\exp\{-\frac{n}{2}[(\theta - z_n)'\Gamma(\theta - z_n)]\}\lambda(d\theta).$$

Theorem 1. *Let conditions* (A1) *to* (A6) *be satisfied. Let $G_{x,n}$ be the gaussian measure just described and let $F_{x,n}$ be the actual conditional distribution of θ given $x = \{x_1, \ldots, x_n\}$. Then*

$$\int \|F_{x,n} - G_{x,n}\| P_{0,n}(dx)$$

tends to zero as $n \to \infty$.

Note 3. The proof is long and will be given in several installments. It should be clear from the proof of Proposition 5, Section 8.3 that the only difficulty is to prove that the $F_{x,n}$ concentrate on balls of the type $\{\theta : |\theta| \leq b/\sqrt{n}\}$. This seems to require a rather involved argument. We have given it in some detail since the proof available in Le Cam [1986, p. 619–621] contains a mistake. (That proof proceeds as if the density f was bounded in a neighborhood of $\theta_0 = 0$.)

Assuming (A1) to (A6), the proof proceeds along the following steps:

Step 1. If (A3) holds for some compact K_0, it also holds if K_0 is replaced by a ball $B(\epsilon) = \{\theta : |\theta| \leq \epsilon\}$ for any fixed $\epsilon > 0$.

Step 2. If there are test functions ω_n with $\int(1 - \omega_n)dP_{0,n} \leq \epsilon_0 < 1/2$ and $\int \omega_n dP_{\tau,n} \leq \epsilon_0 < 1/2$ for $\tau \in B^c(\epsilon)$, then there are numbers $C_1 < \infty$ and $\alpha \in (0, 1)$ and test functions ω'_n such that $\int(1 - \omega'_n)dP_{0,n} \leq C_1\alpha^n$ and $\int \omega'_n dP_{\tau,n} \leq C_1\alpha^n$ for all $\tau \in B^c(\epsilon)$.

Step 3. The prior measure μ can be replaced by its restriction to any fixed ball $B(\epsilon)$.

Step 4. Looking only at the p_τ, $\tau \in K_0$ or $\tau \in B(\epsilon)$, one can proceed as if the observations $x_1, x_2, \ldots x_n$ were real-valued.

Step 5. Assuming that the x_j are real-valued. Let $H_\theta(x) = p_\theta\{(-\infty, x]\}$. Then, there is an $\epsilon_1 > 0$ and a $c > 0$ such that $\sup_x |H_\theta(x) - H_0(x)| \geq c|\theta|$ for all $\theta \in B(\epsilon_1)$.

Step 6. There are test functions ω_n and coefficients C_2 such that $\int(1 - \omega_n)dP_{0,n} \to 0$ and $\int \omega_n dP_{\theta,n} \leq C_2 \exp\{-nc^2|\theta|^2/2\}$ for all $\theta \in B(\epsilon_1)$.

Step 7. If $\{b_n\}$ is a sequence tending to infinity and $C_n = \{\theta : n^{-1/2}b_n \leq |\theta| \leq \epsilon_1\}$ then $\int F_{x,n}(C_n)P_{0,n}(dx) \to 0$.

Step 8. It is sufficient to look at the behavior of posterior distributions for prior measures $\mu_{n,b}$ restrictions of μ to balls $B(n^{-1/2}b)$, b fixed, arbitrary.

Step 9. One can replace the $\mu_{n,b}$ of Step 8 by the restriction to $B(n^{-1/2}b)$ of the Lebesgue measure.

Step 10. To the Lebesgue measure $\lambda_{n,b}$ restricted to $B(n^{-1/2}b)$, one can apply arguments similar to those of Section 6.4.

Actually, Steps 1, 2, 4, 5 and 6 do not involve prior measures at all; they are just statements about tests in the i.i.d. case. They can be found in Le Cam and Schwartz [1960] and Le Cam [1966]. They are also in Le Cam [1986]. We will defer their proofs for the time being.

A connection between the behavior of posterior distributions and that of tests can be obtained from the following easy lemma. It will be stated without any suffixes "n" because it has nothing to do with n.

First note that if C is the set with complement B restricting the measure μ to B yields the inequality

$$\|F_x - F'_x\| \leq 2F_x(C)$$

for F_x a conditional distribution for the prior μ and F'_x a conditional distribution for the restricted measure. For the next statement write $M_B = \int_B P_\theta \mu(d\theta)$ and $P_B = M_B/\mu(B)$. Then note the following inequality.

Lemma 2. *Let U and C be two disjoint measurable subsets of Θ with $\mu(U) > 0$. Then for any probability measure Q on the σ-field $\mathcal{A}$ and any test function ω one has*

$$\begin{aligned}\int F_x(C)Q(dx) &\leq \frac{1}{2}\|P_U - Q\| + \int [1 - \omega(x)]Q(dx) \\ &+ \frac{1}{\mu(U)} \int \omega(x) M_C(dx)\end{aligned}$$

Proof. Write $F_x(C) = [1-\omega(x)]F_x(C)+\omega(x)F_x(C)$. This yields

$$\begin{aligned}\int F_x(C)Q(dx) &\leq \int [1 - \omega(x)]Q(dx) + \int \omega(x)F_x(C)P_U(dx) \\ &+ \int \omega(x)F_x(C)[Q(dx) - P_U(dx)].\end{aligned}$$

The first term on the right was kept as such in the inequality. Since $0 \leq \omega(x)F_x(C) \leq 1$, the third term is seen to be smaller than $\|P_U - Q\|/2$. By the Radon-Nikodym density representation, the second term on the right is $\int \omega(dM_C/dM)dP_U$ where $M = M_\Theta$. One can rewrite it as $\int \omega(dP_U/dM)dM_C$. Since $dP_U/dM \le 1/mu(U)$, the result follows. □

This lemma can be applied with various choices for measure Q. Here we shall use it with $P_{0,n}$ and with $(1/\lambda(U_n)) \int_{U_n} P_{\theta,n}\lambda(d\theta)$ or with $(1/\mu(U_n)) \int_{U_n} P_{\theta,n}\mu(d\theta)$, for choices of neighborhoods U_n that shrink to $\theta_0 = 0$ fairly rapidly. For the study of such measures, the following lemma is useful.

Lemma 3. *Let $U_n(b) = \{\theta; |\theta| \le b/\sqrt{n}\}$. Then the sequences $\{P_{0,n}\}$, $\{(1/\lambda[U_n(b)]) \int_{U_n(b)} P_{\theta,n}\lambda(d\theta)\}$, and $\{(1/\mu[U_n(b)]) \int_{U_n(b)} P_{\theta,n}\mu(d\theta)\}$ are contiguous. In addition the difference*

$$\frac{1}{\lambda[U_n(b)]}\int_{U_n(b)} P_{\theta,n}\lambda(\theta) - \frac{1}{\mu[U_n(b)]}\int_{U_n(b)} P_{\theta,n}\mu(d\theta)$$

tends to zero in L_1-norm.

Proof. For the last statement, just apply condition (A6). For the contiguity statement, note that according to Section 7.2, if $\{\theta_n\}$ is any sequence such that $|\theta_n| \le b/\sqrt{n}$, the sequences $\{P_{0,n}\}$ and $\{P_{\theta_n,n}\}$ are contiguous. This means that $P_{0,n}(A_n) \to 0$ implies that $P_{\theta_n,n}(A_n) \to 0$ and so does an average of the $P_{\theta_n,n}(A_n)$ over $U_n(b)$. The converse is also true. Hence the result. □

Now, assuming that we have already obtained the results described in Step 1 and Step 2, we can apply Lemma 2 for test functions such that $\int \omega_n dP_{\theta,n} \le C_1\alpha^n$ for all $\theta \in B^c(\epsilon)$ and $\int(1-\omega_n)dP_{0,n} \to 0$. In that Lemma 2, put $Q = P_{0,n}$ and $U = U_n(\beta_n) = \{\theta : |\theta| \le \beta_n/\sqrt{n}\}$. Finally put $C = B^c(\epsilon)$. This gives a bound

$$\int F_{x,n}(C)P_{0,n}(dx) \le \frac{1}{2}\|P_{U_n,n} - P_{0,n}\| + \int [1-\omega_n(x)]P_{0,n}(dx)$$

$$+\frac{1}{\mu(U_n)}\int \omega_n(x)M_C(dx).$$

If $\beta_n \to 0$, then $\|P_{U_n,n} - P_{0,n}\|$ tends to zero. The integral $\int \omega_n(x) M_c(dx)$ is bounded by $C_1 \alpha^n$ with $\alpha \in (0, 1)$. The denominator $\mu(U_n)$ is of the order of $(\beta_n/\sqrt{n})^k$. Thus taking $\beta_n \geq 1/\sqrt{n}$ one will still have that $\alpha^n/\mu(U_n) \to 0$.

By the argument about restricting prior measures to a subset $B(\epsilon)$ if $F_{x,n}[B^c(\epsilon)]$ tends to zero, one sees that it will be enough to proceed as if μ was concentrated on a ball $B(\epsilon)$. However, $\epsilon > 0$ is arbitrary, and one can take it so that $0 < \epsilon < \epsilon_1$ of Step 5.

Now assume that we have test functions ω_n as described in Step 6. Take a sequence $\{b_n\}$, $b_n > 0$, $b_n \to \infty$ so that $b_n/\sqrt{n} < \epsilon_1$ and let $W_n(b_n) = \{\theta : b_n/\sqrt{n} \leq |\theta| \leq \epsilon_1\}$.

Lemma 4. *If for $\theta \in B(\epsilon_1) = \{\theta : |\theta| \leq \epsilon_1\}$ the inequality $\int \omega_n dP_{\theta,n} \leq C_2 \exp\{-nc^2|\theta|^2/2\}$ holds and if* (A6) *holds, then, with $U_n = \{\theta : |\theta| \leq 1/\sqrt{n}\}$, the integrals $(1/\mu(U_n)) \int_{W_n(b_n)} [\int \omega_n dP_{\theta,n}]\mu(d\theta)$ tend to zero.*

Proof. It is clear from (A6) that one can replace $\mu(U_n)$ by $\lambda(U_n)$. Doing so, consider first the integral

$$J_n = \frac{1}{\lambda(U_n)} \int_{W_n(b_n)} \exp\{-\frac{nc^2|\theta|^2}{2}\}\lambda(d\theta).$$

Writing $\theta = t/\sqrt{n}$, it becomes

$$J_n = \frac{1}{\lambda(B_1)} \int_{V_n} \exp\{-\frac{c^2|t|^2}{2}\}\lambda(dt)$$

where $B_1 = \{t : |t| \leq 1\}$ and $V_n = \{t : b_n \leq t \leq \epsilon_1\sqrt{n}\}$. Thus J_n tends to zero whenever $b_n \to \infty$.

The measure μ has a density f with respect to λ. For any positive constant A one sees that

$$J'_n = \frac{1}{\lambda(U_n)} \int_{W_n(b_n)} \exp\{-\frac{nc^2|\theta|^2}{2}\}(f(\theta) \wedge A)\lambda(d\theta)$$

will tend to zero. Thus it will be enough to show that the integral where f is replaced by $g = f - (f \wedge A)$ will also tend to zero.

Condition (A6) says that, for θ close to zero, $f(\theta)$ is close to a constant $a > 0$. Take $g = f - [f \wedge (3a)]$. Then, by (A6), integrals of the type $(1/\lambda(B)) \int_B g(\theta)\lambda(d\theta)$ will tend to zero as the radius of the ball B tends to zero. Since for a ball of radius x the measure $\lambda(B)$ is proportional to x^k, this says that $\gamma(x) = x^{-k} \int_{|\theta| \leq x} g(\theta)\lambda(d\theta)$ tends to zero as $x \to 0$.

Let us look at the integrals

$$I_n = \frac{1}{\lambda(U_n)} \int_{W_n(b_n)} \exp\{-\frac{nc^2|\theta|^2}{2}\} g(\theta)\lambda(d\theta).$$

Decompose $W_n(b_n) = \{\theta : b_n/\sqrt{n} \leq |\theta| \leq \epsilon_1\}$ into two pieces $W_n' = \{\theta : b_n/\sqrt{n} \leq |\theta| \leq \beta_n/\sqrt{n}\}$ and $W_n'' = \{\theta : \beta_n/\sqrt{n} \leq |\theta| \leq \epsilon_1\}$ where $\{\beta_n\}$ is an increasing sequence such that $\beta_n/\sqrt{n} \leq \epsilon_1$.

Let us look first at

$$\begin{aligned} I_n'' &= \frac{1}{\lambda(U_n)} \int_{W_n''} \exp\{-\frac{nc^2|\theta|^2}{2}\} g(\theta)\lambda(d\theta) \\ &\leq \frac{1}{\lambda(U_n)} \exp\{-\frac{c^2}{2}\beta_n^2\} \int_{W_n''} g(\theta)\lambda(d\theta). \end{aligned}$$

Since $\lambda(U_n)$ is proportional to $n^{k/2}$ the quantity I_n'' will tend to zero if $n^{k/2}\exp\{-c^2\beta_n^2/2\}$ stays bounded. Thus I_n'' will tend to zero for $\beta_n^2 = (k/c^2)\log n$.

The other integral, I_n', is inferior to a multiple of

$$R_n = n^{k/2} \int_{|\theta| \leq z_n} \exp\{-\frac{nc^2|\theta|^2}{2}\} g(\theta)\lambda(d\theta)$$

where we have written $z_n = \beta_n/\sqrt{n}$ for simplicity. Passing to polar coordinates, this integral becomes

$$R_n = C(k)n^{k/2} \int_0^{z_n} \exp\{-\frac{nc^2\rho^2}{2}\} \varphi(\rho)\rho^{k-1} d\rho$$

where $\varphi(\rho)$ is the average of $g(\theta)$ over the sphere $\{\theta : |\theta| = \rho\}$.

Similarly $\int_{|\theta| \leq x} g(\theta)\lambda(d\theta)$ can be written proportional to the integral $G(x) = \int_0^x \varphi(\rho)\rho^{k-1}d\rho$. Now integrate by parts to obtain

$$R_n = C'(k)n^{k/2} \int_0^{z_n} \exp\{-\frac{nc^2\rho^2}{2}\} dG(\rho)$$

$$= C'(k)n^{k/2}\exp\{-\frac{nc^2\rho^2}{2}\}G(\rho)||_0^{z_n}$$

$$+ \ C'(k)n^{k/2}\int_0^{z_n} G(\rho)\exp\{-\frac{nc^2\rho^2}{2}\}nc^2\rho\, d\rho.$$

The all-integrated term contains a term

$$n^{k/2}\exp\{-\frac{c^2\beta_n^2}{2}\}G(\frac{\beta_n}{\sqrt{n}})$$

with $G(\beta_n/\sqrt{n}) \leq (\beta_n/\sqrt{n})^k\gamma_0(\beta_n/\sqrt{n})$ in which $\gamma_0(x) = \sup_{y\leq x} G(y)y^{-k}$ tends to zero as $x \to 0$. Since $\exp\{-c^2\beta_n^2/2\}\beta_n^k$ stays bounded, this all-integrated term tends to zero.

The remaining integral term is bounded by a constant times

$$\gamma_0\left(\frac{\beta_n}{\sqrt{n}}\right) n^{\frac{k+2}{2}} \int_0^{z_n} \exp\{-\frac{nc^2\rho^2}{2}\}\rho^{k+1}d\rho$$

$$= \gamma_0\left(\frac{\beta_n}{\sqrt{n}}\right)\int_0^{\beta_n} \exp\{-\frac{c^2\rho^2}{2}\}\rho^{k+1}d\rho.$$

Since $\int_0^\infty \exp\{-c^2\rho^2/2\}\rho^{k+1}d\rho$ is finite, this term also tends to zero.

This concludes the proof of Lemma 4. □

Granting the result of Lemma 4, one can apply Lemma 2 again, taking for the set U of that lemma the ball $U_n = \{\theta : |\theta| \leq 1/\sqrt{n}\}$, for the set C the set $W_n(b_n)$ of Lemma 4 and taking for the measure Q the average $P_{U_n,n} = (1/\lambda(U_n))\int_{U_n} P_{\theta,n}\lambda(d\theta)$. Since $\{P_{U_n,n}\}$ and $\{P_{0,n}\}$ are contiguous (Lemma 3), it follows that $F_{x,n}[W_n(b_n)]$ will tend to zero in $P_{U_n,n}$ probability and therefore also in $P_{0,n}$ probability.

This is true for any sequence $\{b_n\}$ such that $b_n \to \infty$. Therefore, given any fixed $\epsilon > 0$, there will exist numbers b and N such that $n \geq N$ implies $\int F_{x,n}[W_n(b)]P_{0,n}(dx) < \epsilon$.

Returning to the proof of Theorem 1, this means that it will be sufficient to look at what happens on balls $U_n(b) = \{\theta : |\theta| \leq b/\sqrt{n}\}$ for fixed large values of b. Indeed, both the true posterior distribution $F_{x,n}$ and the normal approximation $G_{x,n}$ will satisfy

$\int F_{x,n}[W_n(b)]P_{0,n}(dx) < \epsilon$ and $\int G_{x,n}[W_n(b)]P_{0,n}(dx) < \epsilon$ for a large fixed b and n sufficiently large.

The proof of Theorem 1 can then be concluded as in Section 6.4.

This means that Theorem 1 can be considered proved as long as Steps 1, 2, 4, 5 and 6 are acquired. These steps are of interest by themselves. To save space, we shall only sketch their proof. Details can be found in Le Cam and Schwartz [1960] and Le Cam [1966]. They are reproduced in Le Cam [1986, chapter 17].

Step 1 is a consequence of the following remark. On the compact K_0, the map $\theta \rightsquigarrow p_\theta$ is one to one, continuous for the standard topology of K_0, and the weak topology of measures. That is assumed in (A2) and (A4). Thus the inverse map $p_\theta \rightsquigarrow \theta$ is also continuous and even uniformly continuous.

It is not hard to conclude from this that, if one takes experiments $\{p_{\theta,n}; \theta \in K_0\}$, there will exist uniformly consistent estimates T_n of θ; see Le Cam [1986, p. 595]. To get a uniformly consistent test of $\theta_0 = 0$ against $K_0 \backslash B(\epsilon)$ one can take a function $\omega_{n,1}$ equal to 1 if $|T_n| < \epsilon/2$ and to zero if $|T_n| \geq \epsilon/2$. Then one takes $\omega_{n,2} = \omega_n \wedge \omega_{n,1}$ for the ω_n of assumption (A3).

The existence of the ω'_n with the exponential decay of Step 2 is a consequence of very well-known results on sums of independent variables. See, for instance, Hoeffding [1963]. However, the result needed here is very simple. Suppose that ω_N satisfies the condition (A3). Take an integer r. If $rN \leq n < (r+1)N$ divide the $x_1, x_2, \ldots, x_n$ in successive nonoverlapping batches $\{x_1, x_2, \ldots, x_N\}$, $\{x_{N+1}, x_{N+2}, \ldots, x_{2N}\}$ and so forth. On each such batch we can use a replica of ω_N getting functions $\omega_{N,1}, \ldots, \omega_{N,r}$. The test ω'_n can be obtained by accepting or rejecting $\theta_0 = 0$ according to whether $r^{-1}\sum_{j=1}^{r} \omega_{N,j} > 1/2$ or $\leq 1/2$. As $r \to \infty$, this test will have the appropriate exponential behavior. Details can be found in Le Cam [1986, p. 479].

The validity of Step 4 comes from the following remark: Let D be a dense countable subset of K_0. Consider all likelihood ratios $dp_t/(dp_s + dp_t)$ for s and t in K_0. The map $x \rightsquigarrow \{dp_t/(dp_s + dp_t)(x); s \in K_0, t \in K_0\}$ is a sufficient statistic for $\{p_\theta : \theta \in$

$K_0\}$. It takes its values in $[0,1]^{D\times D}$. However, $[0,1]^{D\times D}$ can be mapped in a one-to-one Borel way onto $[0,1]$.

The validity of Step 5 is a matter of calculus. The condition DQM_o asserts the existence of a derivative in quadratic mean for the square roots of densities. This implies the differentiability in the first mean of the densities themselves. Thus, one can use a Taylor expansion (in θ) for the cumulatives $H_\theta(x) - H_0(x)$ around $\theta_0 = 0$. This and the nonsingularity of the derivatives will give a bound as described in Step 5. For details, see Le Cam [1966] or Le Cam [1986, p. 607].

The statement in Step 6 follows from Step 5 by taking the empirical cumulative H_n and rejecting $\theta_0 = 0$ if $\sup_x |H_n(x) - H_0(x)| \geq (c/2)b_n/\sqrt{n}$.

Theorem 1 has been stated using centerings Z_n. It is not hard to see that such centerings can be constructed. One procedure is to take a minimum distance estimate $\hat{\theta}_n$ that almost minimizes $\sup_x |H_n(x) - H_\theta(x)|$ for θ in K_0. Then one performs a one-step operation as in Section 6.3. This does not guarantee any good behavior, except near $\theta_0 = 0$. However, this is the only thing needed here.

In conclusion, except for various details that should be filled in, we can consider that Theorem 1 has been proved. It will usually be applied in a context where no specific θ has been singled out, as we singled out $\theta_0 = 0$ here. Obviously, it will apply to all θ_0 that satisfy conditions (A1) to (A6).

Sections 2, 3 and 4 of this chapter seem to imply that Bayes procedures are usually well behaved asymptotically and that the prior measures used do not matter much. That it is not always so is the subject of the next section.

8.5 Bayes Procedures Behave Miserably

Lest the reader conclude from the preceding sections that Bayes procedures generally behave well, we must point out that this is not really the case.

A first inkling that they may not even be consistent was given by Schwartz in her 1960 thesis. In a parametric case where the true θ is a point in the support of the prior measure μ, the posterior distribution may not concentrate around θ if μ is too "thin" there. Some of the examples published in Schwartz [1965] are not quite correct. The author died before the correction of galley proofs.

Some gross misbehavior was soon pointed out by Freedman [1963]. He considers a parameter set Θ that consists of all probability measures on the integers $\{1, 2, \ldots\}$. This set is a complete separable metric space for the total variation distance. The set $\mathcal{P}$ of all probability measures on Θ can be metrized by the corresponding dual Lipschitz norm. It will also be a complete separable metric space. Let us denote by $P_{\theta,n}$ the distribution of a sequence $\{x_1, \ldots, x_n\}$ of independent observations from a probability measure θ on the integers. Freedman proved several theorems, one of which says the following.

Proposition 6. *The set of pairs $(\theta, \mu) \in \Theta \times \mathcal{P}$ such that*

$$\limsup_n \int F_{x,n}(U) P_{\theta,n}(dx) = 1$$

simultaneously for all open sets $U \subset \Theta$ is the complement of a countable union of closed sets without interior points.

One can restate this in words as follows. In a topological sense, a set that is a countable union of nowhere dense closed sets is a "small" set. The French call it "meager." Thus, except for those priors μ that belong to a meager set, there will be only a meager set of values of θ where the posterior measures do not wander about aimlessly and indefinitely. That will happen even if the support of μ is the entire space Θ. It is due to the fact that, for nearly all $\theta \in \Theta$, the prior measure gives little weight to small neighborhoods of θ.

To be fair to Bayes procedures, one should add that there are particular prior measures μ such that $F_{x,n}(U_\theta) \to 1$ in $P_{\theta,n}$ probability for all neighborhoods U_θ of θ and all θ. They are described in Freedman [1963] and have given rise to a large literature using the particular case of so-called Dirichlet priors. However, if one takes other fairly arbitrary priors, for most of them the wandering about described in Proposition 6 will occur. This is really a most unfortunate form of misbehavior.

There are other forms of misbehavior that look unexpected at first. Diaconis and Freedman [1986] have pointed out the following fact.

Consider probability measures p on the line that are symmetric around zero in the sense that $p(A) = p(-A)$. Let $\mathcal{F}$ be the set of all such measures. Most prior measures μ on $\mathcal{F}$ will exhibit a misbehavior analogous to that of Proposition 6. However, there are priors μ, such as those obtained from Dirichlet priors, such that for every $p \in \mathcal{F}$ the posterior measure will almost surely concentrate around p if one takes independent observations $\{x_1, x_2, \ldots\}$ from p itself.

Take such a measure μ, but now instead of trying to estimate p, let us assume that the observations are as above but shifted by an amount t. That is, they come from p_t equal to p shifted by t. Now we have two parameters $p \in \mathcal{F}$ and $t \in \Re$.

Let us put a good prior μ on $\mathcal{F}$ and take t independently from some measure ν on $\Re$. Then the prior is $\mu \times \nu$ on $\mathcal{F} \times \Re$. If the measure ν is absolutely continuous with respect to the Lebesgue measure λ and has, for instance, a density $f = d\nu/d\lambda$ that is continuous and strictly positive, everywhere, such as a Cauchy density, one would not expect any trouble. After all, if t was fixed, the posterior distribution would concentrate around the true p_t. If p was known, the posterior distribution of t would concentrate around the true value t_0 of t. It would do so nicely according to Theorem 1, Section 8.4.

In spite of this, Diaconis and Freedman [1986] point out that for many pairs (μ, ν) and many choices of p, the posterior distribution of the one-dimensional parameter t will oscillate indefinitely and never concentrate around the true value t_0 of t. This occurs by virtue of a peculiar phenomenon. To ensure consistency for all $p \in \mathcal{F}$ one must scatter μ around. There will then be many p's that have neighbors that are bumpy, symmetric around zero, but with several modes of about equal heights away from zero. The posterior distribution of t will make it oscillate between those modes.

Such examples show that by combining two "tried and true recipes" for μ and ν separately, one gets a $\mu \times \nu$ that creates trouble.

Another example of Diaconis and Freedman [1989] (unpublished) shows that even in simple cases one may have unexpected behavior. Theorem 1 of Section 8.4 uses prior densities f for which $\theta = 0$ is a Lebesgue point in the sense of condition (A6), Section 8.4. Theorem 1 states only a convergence "in $P_{0,n}$ probability" to the gaussian approximation. One could hope for "convergence almost surely." Indeed that will be true in many cases. However, Diaconis and Freedman point out that, even if the density f is bounded and bounded away from zero, almost sure convergence may not take place. They do that for nice measures p_θ where $\theta \in (-1/2, 1/2)$ and where p_θ is the Bernoulli measure giving mass $(1/2) + \theta$ to zero and $(1/2) - \theta$ to unity.

For these, Le Cam [1953] claims almost sure convergence if f is continuous. The Lebesgue point property for a bounded f makes it look close to continuous, but not close enough.

The difficulty does not occur in the proof of Theorem 1, Section 8.4, that $F_{x,n}[B^c(\epsilon_1)]$ will tend to zero for $P_{0,n}$. There the probabilities decrease exponentially fast, yielding easily almost sure convergence. It occurs in the argument of Lemma 4, Section 8.4. Taking the Bernoulli case and centering Z_n that are the observed frequencies minus $1/2$, one sees that Z_n tends to θ almost surely but that $Z_n - \theta$ oscillates according to the law of the iterated logarithm. The posterior distributions concentrate around Z_n, and they can be made to oscillate if f has enough bumps.

In view of the examples of Freedman and of Diaconis and Freedman, one should exert caution in selecting prior distributions and using Bayes procedures.

As Einstein is reported to have said: "Gott ist raffiniert, aber boshaft ist *sie* nicht."

8.6 Historical Remarks

A form of Bayes' theorem was published posthumously by Bayes in 1763. Laplace, who did not seem to know of Bayes' work, proposed it as a principle in 1774. Later on, Laplace [1810a,b] published a form of what we have called the Bernstein–von Mises theorem (Theorem 1, Section 8.4), from the work of Bernstein [1917] and von Mises [1931]. Actually Bernstein's work has little to do with that of Laplace. He considers weak convergence of posterior distributions conditional on the sample average. von Mises worked on parameters in multinomial distributions.

Fisher, whose work [1922, 1925] parallels that of Laplace in more than one way, does not seem to have added results on the behavior of posterior distributions. This may be because he did not view kindly the use of prior distributions and substituted a philosophy based on "fiducial probabilities." These seem to have been introduced as a result of a logically erroneous argument.

Le Cam [1953] revived Laplace's argument. He used the convergence of posterior distributions to normal ones to obtain a sort of asymptotic minimax theorem, and an asymptotic admissibility result for the one-dimensional case. The conditions used there are very strong. A better attack on the subject was that of Schwartz [1965]. The positive results given here are mostly a rewrite of an unpublished paper of Le Cam [1968] that tried to take advantage of Schwartz's approach.

The bad behavior of Bayes procedures in nonparametric situations is from Freedman [1963] and Diaconis and Freedman [1986]. For more information on tail-free measures, see Dubins and Freedman [1967] and the literature on Dirichlet priors, in particular Ferguson [1973], Doksum [1974] and Lo [1984]. We have not taken a stand on Bayesianism as a philosophy. For a spirited defense, see Berger [1985] and Berger and Wolpert [1988]. Contrary to often expressed opinions, Bayes' approach had not disappeared from statistics in the second quarter of the 20th century. It was quite alive in most places, except those that seem to have fallen under the influence of Fisher. It can certainly be used, as shown here, but in practical situations, it should be used with extreme caution.

Bibliography

Aalen, O. (1978). Non parametric estimation of partial transition probabilities in a multiple decrement model, *Ann. Statist.*, *6*, 534–545.

Aalen, O. (1978). Non parametric inference for a family of counting processes, *Ann. Statist.*, *6*, 701–726.

Akaike, H. (1974). A new look at the statistical identification model. *IEEE Trans. Auto. Control, 19,* 716–723.

Aldous, D. (1989) *Probability Approximations via the Poisson Clumping Heuristic.* Springer-Verlag.

Alexiewicz, A. (1950). On the differentiation of vector valued functions, *Studia Math, 11*, 185–196.

Anderson, T.W. (1955). The integral of a symmetric unimodal function over a symmetric convex set and some probability inequalities, *Proc. Amer. Math. Soc.*, *6*, 170–176.

Ando, T. (1966). Contracted projections in L_p spaces, *Pacific. J. Math.*, no. 17, 391–405.

Arak, T.V. and Zaitsev, A. Yu. (1988). Uniform limit theorems for sums of independent random variables, *Proc. Steklov Inst. Math., 174*.

Assouad, P. (1983). Deux remarques sur l'estimation. *C.R. Acad. Sci. Paris Ser. I Math, 296*, 1021–1024.

Barankin, E.W. (1949). Locally best unbiased estimates, *Ann. Math. Stat., 20*, 477–501.

Barbour, A., Holst, L. and Janson, S. (1992). *Poisson Approximation.* Clarendon Press, Oxford.

Basawa, I.V. and Prakasa Rao, B.L.S. (1980). *Statistical Inference for Stochastic Processes*, Academic Press.

Basawa, I.V. and Scott, D.J. (1983). *Asymptotic Optimal Inference for Non-ergodic Models*, Springer-Verlag.

Beran, R. (1974). Asymptotically efficient adaptive rank estimates in location models, *Ann. Statist., 2*, 63–74.

Berge, C. and Ghouila-Houri, A. (1962). *Programmes, Jeux et réseaux de transport*, Dunod.

Berger, J.O. (1985). *Statistical Decision Theory and Bayesian Analysis*, Springer-Verlag.

Berger, J.O. and Wolpert, R.L. (1988). *The Likelihood Principle*, IMS Lecture Notes vol. 6.

Bernstein, S. (1917). *Theory of Probability* (Russian).

Bickel, P., Klaassen, C., Ritov, Y. and Wellner, J. (1993). *Efficient and Adaptive Estimation for Semiparametric Models.* Johns Hopkins Univ. Press, Baltimore.

Billingsley, P. (1968). *Convergence of Probability Measures*, John Wiley & Sons.

Birgé, L. (1983). Approximation dans les espaces métriques et théorie de l'estimation, *Z. Wahrsch. verw. Gebiete, 65*, 181–237.

Birgé, L. (1984). Sur un théorème de minimax et son application aux tests, *Probab. Math. Statist., 3*, no. 2, 259–282.

Birgé, L. (1984). Stabilité et instabilité du risque minimax pour des variables indépendantes équidistribuées, *Ann. Inst. H. Poincaré, (Probab. Statist.) 20*, 201–223.

Blackwell, D. (1951). Comparison of experiments, *Proc. 2nd Berkeley Symp. Math. Stat. Probab., 1*, 93–102.

Blackwell, D. (1953). Equivalent comparisons of experiments, *Ann. Math. Stat., 24*, 265–272.

Blackwell, D. and Girshick, M.A. (1954). *Theory of Games and Statistical Decisions*, John Wiley & Sons.

Bobrov, A.A. (1937) On the relative stability of sums in a particular case. *Mat. Sbornik.* NS *4 (46)* 99–104.

Bochner, S. (1947). Stochastic Processes, *Ann. of Math., 48*, 1014–1061.

Bochner, S. (1955). *Harmonic Analysis and the Theory of Probability*, Calif. Monographs in Math. Science.

Bohnenblust, H.F., Shapley, L.S., and Sherman, S. (1949). Reconnaisance in game theory. Unpublished RAND memorandum R.M. 208, 1–18.

Boll, C. (1955). Comparison of experiments in the infinite case and the use of invariance in establishing sufficiency. Unpublished Ph.D. Thesis, Stanford University.

Bretagnolle, J. and Huber, C. (1979). Estimation des densités, risque minimax, *Z. Wahrsch. verw. Gebiete, 47*, 119–137.

Brown, L.D. (1966). On the admissibilityof invariant estimates of one or more location parameters. *Ann. Math. Stat.*,**37**, 1087–1136.

Chao, M.T. (1967). Non sequential optimal solutions of sequential decision problems, Thesis, Univ. of Calif., Berkeley.

Chernoff, H. (1954). On the distribution of the likelihood ratio, *Ann. Math. Stat. 27*, 573–578.

Chernoff, H. (1956). Large sample theory, parametric case, *Ann. Math. Stat., 27*, 1–22.

Cramér, H. (1946). *Mathematical Methods of Statistics*, Princeton Univ. Press.

Daniels, H.E. (1961). The asymptotic efficiency of a maximum likelihood estimator, *Proc. 4th Berkeley Symp. Math. Stat. Probab., 1*, Univ. of Calif. Press, Berkeley, 151–163.

Daniell, P.J. (1917). A general form of integral. *Ann. Math.* **19,** 279–294.

Daniell, P.J. (1918) Integrals in an infinite number of dimensions. *Ann. Math.* **20,** 281–288.

Darmois, G. (1945). Sur les lois limites de la dispersion de certaines estimations, *Internat. Statist. Rev., 13*, 9–15.

Davies, R. (1985). Asymptotic inference when the amount of information is random, *Proc. Neyman Kiefer Conf., vol. II*, 841–864.

de Moivre, A. (1733). *The Doctrine of Chances*, The 3rd ed. (1756) has been reprinted by Chelsea, New York (1967).

Diaconis, P. and Freedman, D.A. (1986). On the consistency of Bayes estimates, *Ann. Statist., 14*, 1–67.

Diaconis, P. and Freedman, D.A. (1986). On inconsistent Bayes estimates of location, *Ann. Statist., 14*, 68–87.

Diaconis, P. and Freedman, D.A. (1989). Unpublished.

Dieudonné, J. (1941). Sur le théorème de Lebesgue—Nikodym. *Ann. of Math.*, *42*, 547–556.

Dieudonné, J. (1944). Sur le théorème de Lebesgue—Nikodym II, *Bull. Soc. Math. France*, *72*, 193–239.

Dieudonné, J. (1960). *Foundations of Modern Analysis*, Academic Press.

Doksum, K. (1974). Tail free and neutral random probabilities and their posterior distributions, *Ann. Probab.*, *2*, 183–201.

Donoho, D.L. (1990). One-sided inference about functionals of a density, *Ann. Statist.*, 1390–1420.

Donoho, D.L. and Liu, R. (1991). Geometrizing rates of convergence II and III, *Ann. Statist.*, 633–701.

Donoho, D.L., MacGibbon, B. and Liu, R.C. (1990). Minimax risk for hyperrectangles, *Ann. Statist.*, **18**, 1416–1437.

Donoho, D.L. (1997). Renormalizing experiments for nonlinear functionals, in *Festschrift for Lucien Le Cam*, eds. Torgersen, E.N., Pollard, D. and Yang, G.L., Springer, 167–181.

Donoho, D.L., Johnston, I.M., Kerkyacharian, G. and Picard, D. (1997). Universal near minimaxity of wavelet shrinkage, in *Festschrift for Lucien Le Cam*, eds. Torgersen, E.N., Pollard, D. and Yang, G.L., Springer, 183–218.

Donsker, M. (1951). An invariance principle for certain probability limit theorems, *Mem. Amer. Math. Soc.*, *6*, 1–12.

Doob, J.L. (1934). Probability and statistics, *Trans. Amer. Math. Soc.*, *36*, 766–775.

Doob, J.L. (1936). Statistical estimation, *Trans. Amer. Math. Soc.*, *39*, 410–421.

Doob, J.L. (1948). Application of the theory of martingales, *Coll. Int. du C.N.R.S. Paris*, 22–28.

Doob, J.L. (1953). *Stochastic Processes*, Wiley, New York.

Dubins, L.E. and Freedman, D.A. (1967). Random distribution functions, *Proc. 5th Berkeley Symp. Math. Stat. Probab.*, *2*, Univ. of Calif. Press, Berkeley, 183–214.

Dudley, R.M. (1978). Central limit theorems for empirical measures, *Ann. Probab.*, **6**, 899–929.

Dudley, R.M. (1985). An extended Wichura theorem, definition of Donsker class, and weighted empirical distributions, *Probability in Banach spaces*, Springer-Verlag Lecture Notes in Math No. 1153, 141–178.

Dudley, R.M. (1989). *Real Analysis and Probability,* Wadsworth & Brooks/Cole.

Dudley, R.M. (1999). *Uniform Central Limit Theorems,* Cambridge Univ. Press.

Dugué, D. (1936a). Sur le maximum de précision des estimations gaussiennes à la limite. *Comptes Rendus Paris, vol 202,* 193–196.

Dugué, D. (1936b) Sur le maximum de précision des lois limites d'estimation. *Comptes Rendus Paris, vol 202,* 452–454.

Dugué, D. (1937). Application des propriétés de la limite au sens du calcul des probabilités à l'étude de diverses questions d'estimation, *J. Ecole Polytechnique, 3*, 305–374.

Eberlein, W.F. (1949). Abstract ergodic theorems and weak almost periodic functions, *Trans. Amer. Math. Soc.*, 67, 217–240.

Edgeworth, F.Y. (1908). On the probable error of frequency constants, *J. Roy. Statist. Soc. London, 71*, 381–397, 499–512, 651–678.

Edgeworth, F.Y. (1909). On the probable error of frequency constants, *J. Roy. Statist. Soc. London, 72*, 81–90.

Fabian, V. and Hannan, J. (1977). On the Cramér-Rao inequality, *Ann. Statist., 5*, 197–205.

Fabian, V. and Hannan, J. (1982). On estimation and adaptive estimation for LAN families, *Z. Wahrsch. verw. Gebiete, 59*, 459–478.

Fan, J. Q. (1991). On the estimation of quadratic functionals, *Ann. Statist.*, 19, 1273–1294

Feigin, P. (1986). Asymptotic theory of conditional inference for stochastic processes, *Stochastic Process. Appl., 22*, 89–102.

Feller, W. (1968). *An Introduction to Probability Theory and Its Applications*, Vol. II, 2nd. ed., Wiley, New York.

Ferguson, T.S. (1973). A Bayesian analysis of some non parametric problems, *Ann. Statist, 1*, 209–230.

Ferguson, T.S. (1974). Prior distributions in spaces of probability measures, *Ann. Statist.*, *2*, 615–629.

Fisher, R.A. (1922). On the mathematical foundations of theoretical statistics, *Phil. Trans. Roy. Soc. London, Ser A*, *222*, 309–368.

Fisher, R.A. (1925). Theory of statistical estimation, *Proc. Cambridge Phil. Soc.*, *22*, 700–725.

Fisher, R.A. (1956). Statistical methods and scientific induction, *J. Roy. Statist. Soc. Series B*, *17*, 69–78.

Fréchet, M. (1943). Sur l'extension de certaines évaluations statistiques de petits échantillons, *Internat. Statist. Rev.*, *11*, 182–205.

Freedman, D.A. (1963). On the asymptotic behavior of Bayes estimates in the discrete case, *Ann. Math. Stat.*, *34*, 1386–1403.

Freedman, D.A. (1965). On the asymptotic behavior of Bayes' estimates in the discrete case, II, *Ann. Math. Stat.*, *36*, 454–456.

Fremlin, D.H. (1978). Decomposable measure spaces, *Z. Wahrsch. verw. Gebiete*, *45*, 159–167.

Gauss, C.F. (1809). *Theoria motus corporum coelestium in sectionibus conicis solem ambientium.* Hamburg.

Gauss, C.F. (1821). Theoria combinationis observationum erroris minimis obnoxiae, *Commentationes soc. reg. serien. Gottingensis.*

Gelfand, I.M. (1938). Abstrakte functionen und lineare operatoren, *Mat. Sb.*, *4 (46)*, 238–286.

Gnedenko, B.V. (1939). On the theory of limit theorems for sums of independent random variables, (Russian) *Bull. Acad. Sci.*, U.R.S.S.

Gnedenko, B.V. and Kolmogorov, A.N. (1954). *Limit Distributions for Sums of Independent Random Variables*, Addison-Wesley.

Goria, M.N. (1972). Estimation of the location of discontinuities, Thesis, Univ. of Calif., Berkeley.

Greenwood, P.E. and Shiryayev, A.N. (1985). *Contiguity and the Statistical Invariance Principle*, Gordon and Breach.

Grenander, U. (1981) *Abstract Inference.* Wiley & sons.

Hájek, J. (1962). Asymptotically most powerful rank order tests, *Ann. Math. Stat., 33*, 1124–1147.

Hájek, J. (1970). A characterization of limiting distributions of regular estimates, *Z. Wahrsch. verw. Gebiete, 14*, 323–330.

Hájek, J. (1972). Local asymptotic minimax and admissibility in estimation, *Proc. 6th Berkeley Symposium Math. Stat. Prob., 7*, 175–194.

Hájek, J. and Šidák, Z. (1967). *Theory of Rank Tests*, C.S.A.V. Prague and Academic Press.

Hall, P. and Heyde, C.C. (1980). *Martingale Limit Theory and its Application*, Academic Press.

Halmos, P. and Savage, L. (1949). Application of the Radon-Nikodym theorem to the theory of sufficient statistics, *Ann. Math. Stat.*, 20, 225–241.

Halphen, E. (1957). L'analyse intrinséque des distributions de probabilité, *Publ. Inst. Statist., Univ. Paris, 6*, 79–159.

Hansen, O. and Torgersen, E. (1974). Comparison of linear normal experiments. *Ann. Statist., 2* 367–373.

Hauck, W.W. and Donner, A. (1977). Wald's test as applied to hypotheses in logit analysis, *J. Amer. Statist. Assoc., 72*, 851–853. Corrigendum.(1980). *J. Amer. Statist. Soc., 75*, 482.

Hellinger, E. (1909). Neue Begründung der Theorie quadratische formen von Unendlich vielen Veränderlichen, *J. Reine Angew. Math., 136*, 210–271.

Hodges, J.L. Jr. (1952). Personal communications.

Hodges, J.S. (1987). Assessing the accuracy of normal approximations, *J. Amer. Statist. Assoc., 82*, 149–154.

Hoeffding, W. (1963). Probability inequalities for sums of bounded random variables, *J. Amer. Statist. Assoc., 58*, 13–30.

Hoeffding, W. and Wolfowitz, J. (1958). Distinguishability of sets of distributions (the case of independent and identically distributed chance variables), *Ann. Math. Stat., 29*, 700–718.

Hoffman-Jorgensen, J. (1985). *Probability in Banach Spaces, V. Proceedings,* Medford, USA 1984, Springer.

Hoffman-Jorgensen, J. (1991). *Stochastic Processes on Polish*

Spaces. Various Publication Series 39, Aarhus Universitet, Aarhus, Denmark.

Ibragimov, I.A. and Has'minskii, R.Z. (1981). *Statistical Estimation. Asymptotic Theory*, Springer-Verlag.

Jacod, J. and Shiryaev, A.N. (1987). *Limit Theorems for Stochastic Processes*, Springer-Verlag, New York.

James, W. and Stein, C. (1961). Estimation with quadratic loss. *Proc. Fourth Berkley Symp. Math. Statist. Prob.* 1, 311–319.

Janssen, A., Milbrot, H., and Strasser, H. (1985). *Infinitely Divisible Statistical Experiments, Lecture Notes in Statistics, 27*, Springer-Verlag.

Jeganathan, P. (1980). Asymptotic theory of estimation when the limit of the log-likelihood ratios is mixed normal, Ph.D. Thesis, Indian Statistical Institute.

Jeganathan, P. (1982). On the asymptotic theory of estimation when the limit of the loglikelihood is mixed normal, *Sankhyā, Series A, 44*, part 2, 173–212.

Jeganathan, P. (1983). Some asymptotic properties of risk functions when the limit of the experiment is mixed normal, it Sankhyā, Series A., 45, part 1, 66–87.

Jeganathan, P. (1988). Some aspects of asymptotic theory with applications to time series models, Tech. Report No. 166, The Univ. of Michigan.

Jeganathan, P. and Le Cam, L. (1985). On Lévy's martingale central limit theorem, *Sankhyā, Series A, 47*, part 2, 141–155.

Kakutani, S. (1948). On the equivalence of infinite product measures, *Ann. of Math., 49*, 214–224.

Kallenberg, O. (1983). *Random Measures*, Academic Press, New York.

Kholevo, A.S. (1973). A generalization of the Rao-Cramér inequality, *Theory of Probab. and Appl., vol. 18*, 2, 359–362.

Kneser, H. (1952) Sur un théorème fondamental de la théorie des jeux. *C.R. Acad. Sci. Paris, vol. 234,* 2418–2420.

Kolmogorov, A.N. (1931). Eine Verallgemeinerung des Laplace-Liapounoffschen Satzes, *Izv. Akad. Nauk SSSR Ser Mat.*, 959–962.

Kolmogorov, A.N. (1933). Über die Grenzwertsätze der Wahrscheinlichkeitsrechung, *Izv. Akad Nauk SSSR Ser. Fiz. Mat.*, 363–372.

Kolmogorov, A.N. (1956). Deux théorèmes asymptotiques uniformes pour des sommes de variables aléatoires, *Teor. Verojatnost. i. Prim., 1*, 426–436.

Kolmogorov, A.N. and Tichomirov, V.M. (1959). ϵ-entropy and ϵ-capacity of sets in functional spaces, *Uspehi. Mat. Nauk, 14*, 3–86, *Amer. Math. Soc. Transl. Ser. 2, 17*, 277–364.

Komlós, J., Major, P. and Tusnády, G. (1975). An approximation of partial sums of independent R V'-s, and the sample DF.I, *Z. Wahrsch. verw. Gebiete, 32*, 111–131.

Kraft, C.H. (1955). Some conditions for consistency and uniform consistency of statistical procedures, *Univ. California Publ. Statist., 1*, 125–142.

Laplace, P.S. (1810a). Mémoire sur les formules qui sont fonctions de très grands nombres et sur leurs application aux probabilités, *Oeuvres de Laplace, 12*, 301–345.

Laplace, P.S. (1810b). Mémoire sur les intégrales définies et leur application aux probabilités, *Oeuvres de Laplace, 12*, 357–412.

Laplace, P.S. (1820). *Théorie analytique des probabilités*, 3rd ed. Courcier, Paris.

Le Cam, L. (1953). On some asymptotic properties of maximum likelihood estimates and related Bayes' estimates, *Univ. California, Publ. Statist., vol. 1, 11*, 277–330.

Le Cam, L. (1955). An extension of Wald's theory of statistical decision functions. *Ann. Math. Stat., 26*, 69–81.

Le Cam, L. (1956). On the asymptotic theory of estimation and testing hypotheses, *Proc. 3rd Berkeley Symp. Math. Stat. Probab., 1*, 129–156.

Le Cam, L. (1960). Locally asymptotically normal families of distributions, *Univ. California Publ. Statist., 3*, 37–98.

Le Cam, L. (1963). A note on the distribution of sums of independent random variables, *Proc. Nat. Acad. Sciences, 50*, 601–603.

Le Cam, L. (1964). Sufficiency and approximate sufficiency, *Ann. Math. Stat., 35*, 1419–1455.

Le Cam, L. (1965a). A remark on the central limit theorem,

Proc. Nat. Acad. Science, 54, 354–359.

Le Cam, L. (1965b). On the distribution of sums of independent variables, in *Bernoulli-Bayes-Laplace*, Springer-Verlag, New York, 179–202.

Le Cam, L. (1966). Likelihood functions for large numbers of independent observations, *Research Papers in Statistics*, F.N. David, ed. John Wiley & Sons, New York.

Le Cam, L. (1968). Remarks on the Bernsterin–von Mises theorem (unpublished).

Le Cam, L. (1969). *Théorie Asymptotique de la Décision Statistique*, Univ. of Montréal Press.

Le Cam, L. (1970). On the assumptions used to prove asymptotic normality of maximum likelihood estimates, *Ann. Math. Stat., 41*, 802–828.

Le Cam, L. (1972). Limits of experiments, *Proc. 6th Berkeley Symp. on Math. Stat. and Prob., I.* 245–261.

Le Cam, L. (1973). Convergence of estimates under dimensionality restrictions, *Ann. Statist., 1*, 38–53.

Le Cam, L. (1974). *Notes on Asymptotic Methods in Statistical Decision Theory*, Centre de Recherches Mathématiques, U. of Montréal.

Le Cam, L. (1975). On local and global properties in the theory of asymptotic normality of experiments, *Stochastic Processes and Related Topics*, M. Puri, ed. Academic Press, New York, 13–54.

Le Cam, L. (1977). On the asymptotic normality of estimates, *Proc. Symp. to honor J. Neyman*, Warsaw 1974, 203–217.

Le Cam, L. (1979a). A reduction theorem for certain sequential experiments, II, *Ann. Statist., 7*, 847–859.

Le Cam, L. (1979b). On a theorem of J. Hájek. *In Contributions to Statistics: J. Hájek Memorial Volume* (Jurečkova, J. ed.) Akademia Prague 119–137.

Le Cam, L. (1985). Sur l'approximation de familles de mesures par des familles gaussiennes, *Ann. Inst. H. Poincaré Probab. Statist., 21*, 225–287.

Le Cam, L. (1986). *Asymptotic Methods in Statistical Decision Theory*, Springer-Verlag, New York.

Le Cam, L. (1994). An infinite dimensional convolution theorem, In *Statistical decision theory and related topics, V*, Gupta, S.S. and Berger, J.O., ed. Springer, New York.

Le Cam, L. and Schwartz, L. (1960). A necessary and sufficient condition for the existence of consistent estimates, *Ann. Math. Stat., 31*, 140–150.

Le Cam, L. and Yang, G.L. (1988). On the preservation of local asymptotic normality under information loss, *Ann. Statist., 16*, 483–520.

Legendre, A.M. (1805). *Nouvelles Méthodes pour la Détermination des Orbites des Cométes*, Paris.

Lehmann, E.L. (1949). Some comments on large sample tests, *Berkeley Symp. on Statist. and Prob.* , Univ. of Calif. Press, Berkeley, pp. 451–457.

Lehmann, E.L. (1983). *Theory of Point Estimation*, John Wiley & Sons, New York.

Lehmann, E.L. (1988). Comparing location experiments, *Ann. Statist., 16*, 521–533.

Levit, B. (1973). On optimality of some statistical estimates, *Proc. Prague Symp. Asymptotic Statistics, 2*, 215–238.

Lévy, P. (1937). *Théorie de L'addition des Variables Aléatoires*, Gauthier Villars, Paris.

Lindae, D. (1972). Distributions of likelihood ratios and convergence of experiments, Thesis, Univ. of Calif., Berkeley.

Lindeberg, J.W. (1920). Über das Exponentialgesetz in der Wahrscheinlichkeitsrechnung. *Ann. Acad. Sci. Fennicae, 16*, 1–23

Liptser, R. and Shiryaev, A. (1977). *Statistics of Random Processes, I*, Springer-Verlag.

Liptser, R. and Shiryaev, A. (1978). *Statistics of Random Processes, II*, Springer-Verlag.

Lo, A.Y. (1984). On a class of Bayesian non parametric estimates: I. Density estimates, *Ann. Statist., 12*, 351–357.

Loève, M. (1957). A l'interieur du problème limite central, *Publ. Inst. Statist. Univ. Paris, 6*, 313–326.

Loève, M. (1977). *Probability Theory*, 4th edition, Vol. 1, Springer-Verlag.

Loomis, L.H. (1953). *An Introduction to Abstract Harmonic Analysis.* Van Nostrand, New York.

Maharam, D. (1942). On homogeneous measure algebras, *Proc. Nat. Acad. Sci.* USA. 28, 108–111.

Mammen, E. (1987). Optimal local Gaussian approximation of an exponential family, *Probab. Theory Related Fields, 76*, no. 1, 103–119.

Matusita, K. (1955). Decision rules based on the distance, for problems of fit, two samples, and estimation, *Ann. Math. Stat., 26*, 613–640.

Matusita, K. (1961). Interval estimation based on the notion of affinity, *Bull. Inter. Stat. Inst., 38*, Part 4, 241–244.

Matusita, K. (1967). Classification based on distance in multivariate Gaussian cases, *Proc. 5th Berkeley Symp. Math. Stat. Probab., Vol 1*, 299–304.

McLeish, D. (1975). A maximal inequality and dependent strong laws, *Ann. Prob., 3*, 829–839.

Millar, P.W. (1983). The minimax principle in asymptotic theory, *École d'Été de Probabilités de Saint-Flour* XI-1981. P.L. Hennequin, ed. Lecture Notes in Mathematics, *Vol. 976,* Springer-Verlag, 76–267.

Millar, P.W. (1985). Nonparametric applications of an infinite dimensional convolution theorem, *Z. Wahrsch. verw. Gebiete, 68*, 545–556.

Morse, N. and Sacksteder, R. (1966). Statistical isomorphism, *Ann. Math. Stat., 37*, 203–214.

Moussatat, W. (1976). On the asymptotic theory of statistical experiments and some of its applications, Thesis, Univ. of Calif., Berkeley.

Neyman, J. (1949). Contribution to the theory of the χ^2 test, *Proc. Berkeley Symp. Math. Stat. Probab.*, 239–273.

Neyman, J. (1952). *Lectures and Conferences on Mathematical Statistics and Probablity*, Graduate School, U.S. Dept. of Agriculture, Washington, D.C.

Neyman, J. (1959). Optimal asymptotic tests of composite hypotheses, *The Harald Cramér volume* (Grenander, U. ed.) Almquist and Wiksell, 213–234.

Norberg, E. (1999). Approximate comparison of experiments with filtered probability spaces. Preprint Univ. Oslo.

Nussbaum, M. (1996). Asymptotic equivalence of density estimation and Gaussian white noise. *Ann. Stat.*, *24*, 2399–2430.

Pfanzagl, J. and Wefelmeyer, W. (1982). *Contributions to a General Asymptotic Statistical Theory.*, Lecture Notes in Statistics, *13*, Springer-Verlag.

Pitman, E.J.G. (1979). *Some Basic Theory for Statistical Inference*, John Wiley & Sons, New York.

Prakasa Rao, B.L.S. (1968). Estimation of the location of the cusp of a continuous density, *Ann. Math. Stat.*, *39*, 76–87.

Prakasa Rao, B.L.S. (1987). *Asymptotic Theory of Statistical Inference*, John Wiley & Sons, New York.

Raikov (1938). On a connection between the central limit theorem in the theory of probability and the law of large numbers, *Ivz. Akad. Nauk. Ser. Mat.*, *14*, 323–338.

Rao, C. R., (1945). Information and accuracy attainable in the estimation of statistical parameters, *Bull. Calcutta Math. Soc.*, *37*, 81–91.

Rebolledo (1980). Central limit theorems for local martingales, Z. Wahrsch. verw. Geb., 51, 269–286.

Rényi, A. (1958). On mixing sequences of sets, *Acta. Math. Hungar.*, *9*, 215–228.

Riesz, F. (1940). Sur quelques notions fondamentales dans la théorie générale des opérateurs linéaires, *Ann. of Math.*, *41*, 174–206.

Roussas, G.G. (1972). *Contiguous Probability Measures: Some Applications in Statistics*, Cambridge Univ. Press.

Saks, S. (1937). *Theory of the Integral*, 2nd ed. Hafner, Dover reprint (1964).

Schwarz, G. (1978). Estimating the dimension of a model, *Ann. Stat. 6*, 461–464.

Schwartz, L. (1965). On Bayes procedures, *Z. Wahrsch. verw. Gebiete, 4*, 10–26.

Serfling, R.I. (1980). *Approximation Theorems of Mathematical Statistics*, John Wiley & Sons, New York.

Shiryaev, A.N. (1981). Martingales: Recent developments, results and applications, *Internat. Statist. Review, 49*, 199–233.

Sion, M. (1958). On general minimax theorems, *Pacific J. Math., 8*, 171–176.

Stein, C. (1945). Two sample test of a linear hypothesis whose power is independent of the variance, *Ann. Math. Stat., 16*, 243–258.

Stein, C. (1951). Notes on the comparison of experiments, Univ. of Chicago.

Stein, C. (1956). Efficient nonparametric testing and estimation, *Proc. 3rd Berkeley Symp. Math. Statist. Probab., 1*, Univ. of Calif. Press, Berkeley, 187–195.

Stone, C.J. (1975). Adaptive maximum likelihood estimators of a location parameter, *Ann. Statist., 3*, 267–284.

Strassen, V. (1965). The existence of probability measures with given marginals, *Ann. Math. Stat., 36*, 423–439.

Strasser, H. (1985). *Mathematical Theory of Statistics*, Walter de Gruyter, Berlin-New York.

Swensen, A. (1980). Asymptotic inference for a class of stochastic processes, Ph.D. Thesis, Univ. of Calif., Berkeley.

Torgersen, E.N. (1970). Comparison of experiments when the parameter space is finite, *Z. Wahrsch. verw. Gebiete, 16*, 219–249.

Torgersen, E.N. (1972). Comparison of translation experiments. *Ann. Math. Statist., 43*, 1383–1399.

Torgersen, E.N. (1974). Comparison of experiments by factorization. *Stat. Res. Report*, Univ. of Oslo.

Torgersen, E.N. (1991). *Comparison of Statistical Experiments*, Cambridge University Press.

Vaeth, M. (1985). On the use of Wald's test in exponential families, *Internat. Statist. Rev., 53*, 199–214.

van de Geer, S. (2000). *Empirical Processes and M-Estimation.* Cambridge University Press.

van der Vaart, A.W. (1988). Statistical estimation in large parameter spaces, *C.W.I. Tract #44, Amsterdam.*

van der Vaart, A.W. (1991). On differentiable functionals. *Ann. Statist., 19*, 178–204.

van der Vaart, A.W. (1998). *Asymptotic Statistics*, Cambridge Univ. Press.

von Mises, R. (1931). *Wahrscheinlichkeitsrechnung*, Springer-Verlag.

Wald, A. (1938). Contributions to the theory of statistical estimation and testing hypotheses, *Ann. Math. Stat., 10*, 299–326.

Wald, A. (1943). Tests of statistical hypotheses concerning several parameters when the number of observations is large, *Trans. Amer. Math. Soc., 54*, 426–482.

Wald, A. (1947). *Sequential Analysis*, John Wiley & Sons.

Wald, A. (1950). *Statistical Decision Functions*, John Wiley & Sons.

Wendel, J.G. (1952). Left centralisers and isomorphisms of group algebras. *Pacific J. Math., 2*, 251–261.

Wilks, S.S. (1938). Shortest average confidence intervals from large samples, *Ann. Math. Stat., 9*, 166–175.

Wittenberg, H. (1964). Limiting distributions of random sums of independent random variables, *Z. Wahrsch. verw. Gebiete, 3*, 7–18.

Yang, G.L. (1968). Contagion in stochastic models for epidemics, *Ann. Math. Stat., 39*, 1863–1889.

Yang, G.L. (1997). Le Cam's procedure and sodium channel experiments, in *Festschrift for Le Cam*, eds. Torgersen, E.N., Pollard, D. and Yang, G.L., Springer Springer, 411–421.

Yatracos, Y. (1985). Rates of convergence of minimum distance estimators and Kolmogorov's entropy, *Ann. Statist., 13*, 768–774.

Author Index

Subject Index

Springer Series in Statistics

(continued from p. ii)

Ramsay/Silverman: Functional Data Analysis.
Rao/Toutenburg: Linear Models: Least Squares and Alternatives.
Read/Cressie: Goodness-of-Fit Statistics for Discrete Multivariate Data.
Reinsel: Elements of Multivariate Time Series Analysis, 2nd edition.
Reiss: A Course on Point Processes.
Reiss: Approximate Distributions of Order Statistics: With Applications to Non-parametric Statistics.
Rieder: Robust Asymptotic Statistics.
Rosenbaum: Observational Studies.
Rosenblatt: Gaussian and Non-Gaussian Linear Time Series and Random Fields.
Särndal/Swensson/Wretman: Model Assisted Survey Sampling.
Schervish: Theory of Statistics.
Shao/Tu: The Jackknife and Bootstrap.
Siegmund: Sequential Analysis: Tests and Confidence Intervals.
Simonoff: Smoothing Methods in Statistics.
Singpurwalla and Wilson: Statistical Methods in Software Engineering: Reliability and Risk.
Small: The Statistical Theory of Shape.
Sprott: Statistical Inference in Science.
Stein: Interpolation of Spatial Data: Some Theory for Kriging.
Taniguchi/Kakizawa: Asymptotic Theory of Statistical Inference for Time Series.
Tanner: Tools for Statistical Inference: Methods for the Exploration of Posterior Distributions and Likelihood Functions, 3rd edition.
Tong: The Multivariate Normal Distribution.
van der Vaart/Wellner: Weak Convergence and Empirical Processes: With Applications to Statistics.
Verbeke/Molenberghs: Linear Mixed Models for Longitudinal Data.
Weerahandi: Exact Statistical Methods for Data Analysis.
West/Harrison: Bayesian Forecasting and Dynamic Models, 2nd edition.

Printed in the United States
86524LV00003B/22-30/A

9 780387 950365